Metal Oxide Dielectrics: Overcoming Challenges for Large-Area Electronics

Feddy

Table of contents

Metal oxide based high-k dielectrics and semiconductors (MOS) have gained tremendous attention due to their electronic properties and applicability for various electronic and optoelectronic applications.[1] Over the last decades, transparent metal oxide based thin-film transistors (TFTs) have been extensively studied, especially due to their potential employment in large-area flat-panel display (FPD) applications.[2] In order to achieve high performance TFTs, which are suitable for next-generation flat-panel displays, the optimization of both main components in such TFT devices, the semiconductor as well as dielectric material, is of highest importance. Although significant progress has already been achieved with regard to metal oxide semiconductors, research concerning the application of novel metal oxide dielectrics is still highly beneficial in order to improve the current situation.

Thereby, various binary metal oxide high-k dielectrics such as Ta_2O_5,[3] HfO_2,[4] ZrO_2,[5-7] Y_2O_3 [8-12] and Al_2O_3 [2, 12-15] have shown promising results regarding their dielectric properties in such electronic devices and thus could potentially replace the established SiO_2 gate dielectric in thin-film transistors. However, these high-k materials are mainly deposited by expensive vacuum-based techniques, which create challenges regarding large-area fabrications.[16] Consequently, extensive research has been dedicated to the solution-processing of functional metal oxide thin films, in order to realize a cost-efficient, large-area deposition of these fundamental building blocks for electronic applications.[16, 17] Thereby, the solution-processing technique is frequently performed by employing metal salt precursors, e.g. alkoxides, acetates, chlorides or nitrates, which exhibit an easy commercial accessibility and low-cost. However, intrinsic precursor properties and the requirement for high processing temperatures as well as stabilizers and other additives are still limiting the industrial application regarding large-area fabrications.[18, 19] The use of stabilizers and other additives can cause an inconsistency in the precursor conversion as well as in the resultant metal oxide thin film formation.

Hence, the employment of single-source molecular precursors, wherein all the necessary chemical components are integrated into one defined molecule, could certainly help to increase the degree of reproducibility, in order to enable the desired decomposition into a high quality metal oxide. Such single-source molecular precursors can be purposely tailored, via coordination chemistry, to possess superior chemical and physical properties, which enable consistent processing of high performance functional metal oxide thin films on a large scale.

The objective of this book consists of the development of various solution-processing techniques for the formation of metal oxide high-k dielectric thin films of aluminum oxide (Al_xO_y), yttrium oxide (Y_xO_y) and the compositional ternary hybrid yttrium aluminum oxide (YAl_xO_y). Thereby, the synthesis of the high-k dielectric thin films as well as their integration in capacitor and thin-film transistor devices is achieved for the first time by a single-source molecular precursor (SSP) route.

Additionally, the synthesis and complete structural characterization of novel single-source molecular precursors of tin, indium, gallium and zinc is accomplished in the course of this book, which are suitable for the formation of electronically active, high performing multinary metal oxide semiconductors, such as zinc tin oxide (ZTO) and indium gallium zinc oxide (IGZO).

The first chapters of the present cumulative book emphasizes on the importance of metal oxide high-k dielectrics and semiconductors for TFT applications as well as the technological advances by employing single-source molecular precursors for the formation of functional metal oxide thin films. The following chapters present a brief overview regarding the fundamentals of binary and multinary metal oxide semiconductors (chapter 2) and high-k dielectrics (chapter 3). Subsequently, the theoretical background and functionality of capacitors and thin-film transistors is discussed. The chapters 4 and 5 present advanced solution-processing techniques, including combustion synthesis of metal oxide high-k dielectric thin-films in organic and aqueous solvents as well as photo-processing of such metal oxide thin films. In addition, the material and electrical properties of the functional metal oxides derived from the novel single-source molecular precursors are characterized and physical as well as chemical structure-property relations are established. Thereafter, the arrangement of the cumulative book, based on the published results, is provided in chapter 6.

1.1 The importance of metal oxide high-k dielectrics and semiconductors for thin-film transistor applications

Thin-film transistors represent the most important component in integrated electronic circuits (ICs)[20, 21], which can be employed for a wide range of applications, such as X-ray detectors[22, 23], chemical and biochemical sensors[24, 25], photovoltaics or portable electronics.[26] In fact, ICs are fundamental building blocks in practically every modern-day electronic device, e.g. TVs, computers, notebooks, smart phones and especially flat-panel displays (FPD).[27-30] Regarding the FPD industry, TFTs are primarily used as backplane electronics for active-matrix liquid-crystal displays (AMLCDs) and active-matrix organic light-emitting diode displays (AMOLEDs), functioning as switch and allowing to drive each pixel of an image independently.[31] Thereby, each pixel consists of three sub-pixels and each sub-pixel is individually connected to a TFT as well as a charge storage capacitor.[21, 32] In its most simplest form, TFT devices consist of three electrodes called "gate", "source" and "drain" and moreover the dielectric and semiconductor layers, which represent the two main components in such TFT devices (Figure 1).

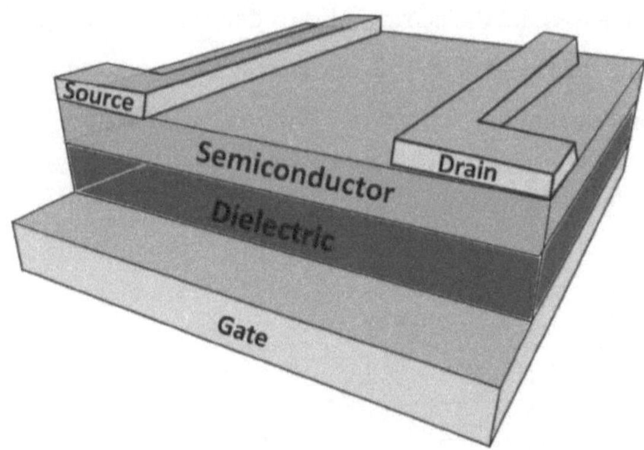

Figure 1 Schematic illustration of a thin-film transistor device.

Thin-film transistors are operating under an electric-field and thus are categorized as field-effect transistor (FET). The functionality of the device is based on the ability to control the current flow through the semiconductor layer between the source and the drain electrodes, induced by the generated electric-field between the gate and drain electrodes. The dielectric layer is incorporated between the semiconductor layer and gate electrode, in order to ensure that no current flows between the gate and the drain electrode and thus preventing the occurrence of electrical short circuits. The intrinsic properties of the active semiconductor layer directly determine the amount of available charge carriers, whereby the flow of charge carriers in TFTs is determined by the accumulation of charges at the semiconductor/dielectric interface. Besides, the semiconductor/dielectric interface properties have a crucial impact on the overall TFT performance characteristics and thus the optimization of the semiconductor as well as dielectric material is of highest importance, in order to ensure further TFT performance enhancements in an efficient manner. In the last decades, the microelectronic industry was primarily based on silicon/silicon dioxide (Si/SiO_2) regarding the semiconductor/dielectric material combination in thin-film transistor devices, which have almost reached their fundamental material limits nowadays.[33, 34]

SiO_2 exists as an excellent amorphous dielectric material with very few electronic defects, forming an excellent interface with Si and typically exhibiting electrical breakdown fields above 10 MV cm^{-1}.[32, 35, 36] Furthermore, silicon dioxide can be etched and is patternable to a nanoscopic scale.[36] Unfortunately, SiO_2 exhibits one significant drawback that cannot be ignored - SiO_2 possess a relatively small dielectric constant of $k = 3.9$. Therefore, downscaling to very thin layers is required (d < 2 nm) in order to further enhance the capacitance, resulting in quantum tunnel phenomena across the dielectric and thus deteriorating the final device performance.[37] In fact, a SiO_2 film thickness of d < 2 nm is reaching an unacceptable power dissipation, exhibiting a leakage current density of J > 1 A cm^{-2} at 1 V, (Figure 2).[38]

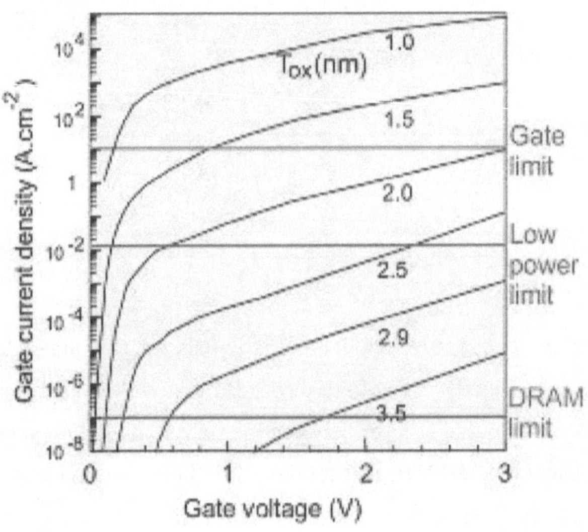

Figure 2 Leakage current density vs. voltage for various layer thicknesses of SiO_2. Measurements from[38]. Adapted with permission[34]. Copyright 2014, Elsevier B.V.

As a consequence, for future TFT performance improvements, it is of immense importance to focus on materials with enhanced dielectric constants, so called "high-k" dielectrics, e.g. Ta_2O_5, HfO_2, ZrO_2, Y_2O_3 and Al_2O_3. These materials enable low-voltage operations as well as the employment of thicker dielectric layers, resulting in lower leakage current densities while still exhibiting sufficiently high capacities.[17] As a result, thin-film transistors utilizing high-k dielectrics exhibit higher mobilities and lower threshold voltages in comparison to TFTs using the established SiO_2 dielectric.[39, 40] Thus the synthesis of high-k materials as a replacement for the conventional SiO_2 as gate dielectric is inevitable, due to the industrial requirements of future device miniaturization combined with the demand for high performance TFTs.

Besides, elemental silicon (Si) itself exhibits merely average semiconductor performance characteristics compared to various metal oxide semiconductors, which had been extensively investigated over the last decades.[27, 41] Thus, the industrial use of Si as active channel material might be plainly due to the fact that SiO_2 can simply be grown by the thermal oxidation of Si, resulting in excellent semiconductor/dielectric interface properties. The moment the SiO_2 dielectric is replaced, Si loses its main advantage as active semiconductor and thus the MOS can be considered as choice for the use as active channel material.[42] However, in order to achieve high performance TFTs compatible with ever increasing electronic demands, it is inevitable to focus on both the critical components of such TFT devices - the semiconductor and the equally essential dielectric material.

1.2 Single-source molecular precursors in solution-processing of metal oxide semiconductor and high-*k* dielectric thin films

Until now, most high-performance metal oxide based TFTs are manufactured by utilizing costly vacuum-based deposition techniques, such as atomic layer deposition (ALD)[43-48], chemical vapor deposition (CVD)[49-52] as well as various sputter deposition techniques.[53-59] As a consequence, intense research has been dedicated towards the fabrication of TFTs fabricated by an alternative deposition technique, which is based on the solution-processing of functional metal oxide thin films.[15] Thereby, the most commonly approached deposition techniques are chemical bath deposition (CBD)[60-62], spin-coating (SC)[9, 11, 13, 18, 63-65] and inkjet printing.[66-71] Regarding the synthesis of such functional metal oxide thin films, via the solution-processing method, the major part of publications report the employment of metal salt precursors, such as nitrates, chlorides, carboxylates and alkoxides, mainly due to their commercial availability and low-cost. However, most of them possess significant drawbacks, such as the requirement of high processing temperatures in order to achieve a clean final product as well as the need for stabilizers and other additives, which is still restricting the industrial application of large-area fabrications.[18, 19] The employment of such additives can have an uncertain impact on the thin film formation and thus the resulting electrical properties[72].

In order to overcome such existing processing obstacles for the solution-processing of functional metal oxide thin films, single-source molecular precursors, with superior chemical and physical properties, are employed in the course of this book. Single-source molecular precursors represent coordination compounds which feature predefined metal-oxygen (M-O) bonds and thus allow for the design of a molecular template, resembling the arrangement and connectivity of the atoms which are present in the final metal oxide on a molecular level. Furthermore, this approach enables the possibility of customizing the precursor ligand structure in order to obtain the desired chemical and physical properties, which is essential for the achievement of high quality metal oxide thin films. The desirable properties of the molecular precursors include air-stability, high solubility in aqueous and organic solvents, decreased post-deposition annealing (PDA) temperatures, a clean decomposition and the applicability of direct photo-patterning. As a result, such well-defined single-source molecular precursors are highly promising materials for an efficient large-area solution-processing of functional metal oxides. In the year 2008, our group first reported on the low-temperature solution-processing of transparent, semiconducting ZnO thin films and their electronic performance in TFT devices, by employing the molecular precursor compound bis[2-(methoxyimino)propanoato]zinc, referred to as "oximato" complex (Figure 3a).[73]

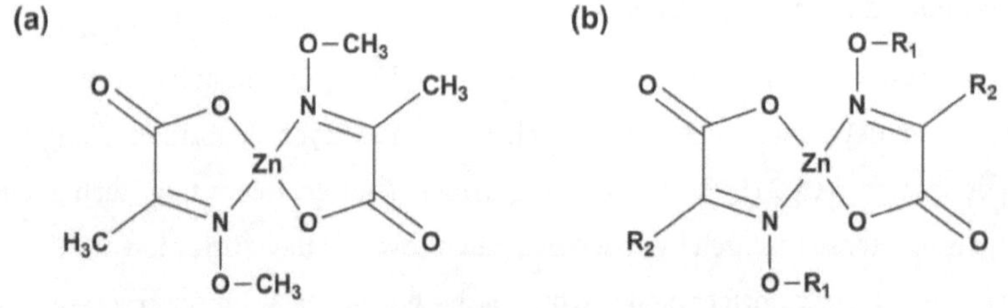

Figure 3 a) Structure of bis[2-(methoxyimino)propanoato]zinc, b) general structure of the zinc precursor with substitutable residues R1 and R2.

This study clearly demonstrated the ability to control the physical properties of the molecular precursors by undergoing defined changes in the molecular structure, whereby different residues R1 and R2 were synthesized by a Schiff Base condensation (Figure 3b). As a result, the processing properties are significantly more influenced by residue R1 in comparison to residue R2. When R1=H, the molecular precursors require higher decomposition temperatures and possess a low solubility in various common solvents, while solutions of compounds with R1=C_2H_5 show a poor wettability on various substrates. The employment of precursors with R1 = CH_3 (Figure 3a) enables the formation of zinc oxide thin films with excellent adhesion on various substrates. Subsequently, it was possible to fabricate TFTs clearly demonstrating the feasibility of the solution-processed ZnO semiconductor, derived from a single-source molecular precursor route. In fact, the executed solution-processing technique represents a highly promising deposition method towards low-cost and large-area fabrication of functional metal oxide thin films.[20, 74, 75] The solution-processing approach is based on precursor chemistry, whereby various precursor compounds can be employed, which are dissolved in aqueous or organic solvents prior to the thin film deposition, e.g. via the spin-coating technique. Subsequently, the deposited precursor thin film gets converted into the desired metal oxide, by means of thermal annealing (see Figure 4) or e.g. by means of photoactivation. Thereby, the employment of single-source molecular precursor compounds enable solution-processing of functional metal oxide thin films without the requirement of additives or stabilizing agents, even under an inert atmosphere. Figure 4 illustrates the individual processing steps involved in the solution-processing of functional metal oxides by employing single-source molecular precursors. Moreover, these molecular precursor compounds undergo a defined decomposition mechanism under thermal treatment, resulting in a clean final product by releasing volatile by-products, and thus allowing a more accurate control of the respective metal oxide thin-film formation. Figure 4, right side demonstrates the thermal decomposition pathways of the single-source molecular precursor compounds bis[(methoxyimino)-propanoato]zinc(II) and bis[(ethoxyimino)-propanoato]zinc(II).[76]

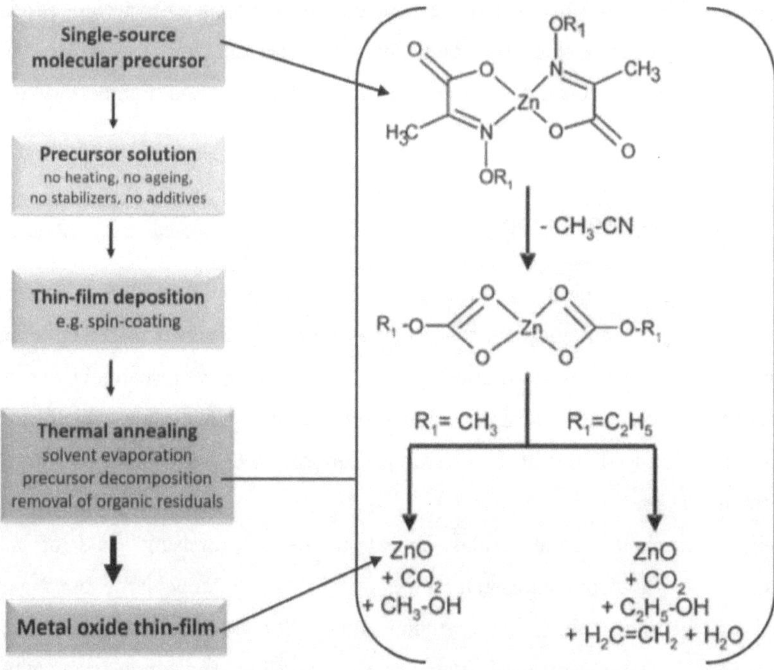

Figure 4 Schematic illustration of individual process steps involved in the solution-processing of metal oxide thin films employing single-source molecular precursors; (right side) decomposition mechanism of zinc oximato complexes, determined by TG-MS and TG-IR. Adapted with permission[76]. Copyright 2009, Royal Society of Chemistry.

Additionally, solution-processing enables a simple approach to generate multinary metal oxide thin films, by dissolving the respective metal oxide precursors in the desired ratio prior to the thin film deposition, which allows the fine-tuning of the electrical properties of the compositional hybrid metal oxides. Hence, in later works performed by our group, the fabrication of high performance TFTs was accomplished by the synthesis of multinary semiconducting indium zinc oxide (IZO) thin films, introducing an indium coordination compound, possessing 2-methoxyimino-propionic ligands, analogues to the previously established zinc "oximato" precursor.[77] Furthermore, our group successfully transferred the SSP approach to the generation of metal oxide high-k dielectric thin films of zirconium oxide (Zr_xO_y), hafnium oxide (Hf_xO_y) and tantalum oxide (Ta_xO_y).[78-80]

In the course of this book, such a highly promising single-source molecular precursor approach is further extended to the generation of semiconducting indium-free zinc tin oxide (ZTO) and quaternary indium gallium zinc oxide (IGZO) as well as amorphous metal oxide high-k dielectrics of aluminum oxide (Al_xO_y), yttrium oxide (Y_xO_y) and yttrium aluminum oxide (YAl_xO_y). In the process, several molecular precursor compounds are synthesized, structurally elucidated and advanced processing techniques are developed, in order to accomplish the solution-processing of metal oxide based high performance capacitor and TFT devices.

2 Metal oxide semiconductors

Semiconductors and dielectrics represent the main components in TFT devices, whereby the semiconductor is functioning as active channel material, which is in direct contact with the dielectric and thus forming an interface with it. In fact, extensive research were executed in the last decades, in order to determine suitable semiconductor materials, which can surpass the performance characteristics of the established amorphous Si semiconductor, suitable for high performance TFTs.[20]

Thereby, amorphous metal oxides appear to be a promising class of materials for next-generation semiconductor-technology, due to their relative large band gaps (E_g) and optical transparency, which makes them suitable for transparent electronics and optical displays.[1, 81-86] The progress of such amorphous oxide semiconductors (AOS) as active component in TFT devices will be discussed shortly in the following.

2.1 Binary metal oxide semiconductors

Binary metal oxide semiconductors such as tin dioxide (SnO_2), indium oxide (In_2O_3) and zinc oxide (ZnO) are studied extensively over the last decades and are well established components as active channel layer material in TFT devices. Unfortunately, all of them possess significant drawbacks, which renders the technological suitability of these binary AOS complicated.

Tin (IV) oxide (SnO_2) was the first metal oxide which was employed as active channel material in TFTs in the 1960s and was therefore intensively investigated by many research groups. Nevertheless, the requirement of higher annealing temperatures in combination with the need of harsh etching conditions renders SnO_2 less suitable as semiconductor material from a technological point of view.[87, 88]

Indium oxide (In_2O_3) is a transparent, highly conducting oxide possessing a high intrinsic charge carrier concentration ($> 10^{19}$ cm^{-3}) and thus is frequently used as electrode material.[89] Therefore, it is necessary to reduce the overall charge carrier concentration, in order to reach the semiconducting magnitude of $< 10^{18}$ cm^{-3}, allowing the use of In_2O_3 as an efficient active channel material in TFT devices.

Zinc oxide (ZnO) possess a drastically decreased intrinsic charge carrier concentration (10^{15} - 10^{17}) in comparison to In_2O_3.[1] Additionally, the deposition process of high-performing ZnO based TFTs requires a controlled atmosphere with certain oxygen partial pressure[90], which was optimized by the Fortunato group reaching mobilities of up to $\mu \approx 50$ cm^2 V^{-1} s^{-1}.[91]

Similar optimization progresses regarding the fabrication of In_2O_3 and SnO_2 based TFTs could also be achieved over time, but due to the inhomogeneous deposition of these binary metal oxides on large-area substrates, they currently not accomplish technological demands. The inhomogeneous thin-film formation can be attributed to the polycrystalline nature of these binary oxides, rendering the reproducibility complicated and resulting in poor TFT performances.[92] As a consequence, investigations of amorphous multinary metal oxide semiconductors became attractive, due to the ability of tuning the electrical properties by combining these metal oxides.

2.2 Multinary metal oxide semiconductors

Among the multinary metal oxide semiconductors hybrids of ZnO, In_2O_3, SnO_2 and Ga_2O_3 are the most prominent choices, namely the ternary semiconducting oxides indium zinc oxide (IZO) and zinc tin oxide (ZTO) as well as the quaternary semiconducting metal oxide indium gallium zinc oxide (IGZO). Due to the feasibility of these multinary metal oxides for obtaining high performance TFTs, they were particularly investigated in detail. The following chart demonstrates the research distribution in the field of metal oxide semiconductors in recent years (Figure 5).

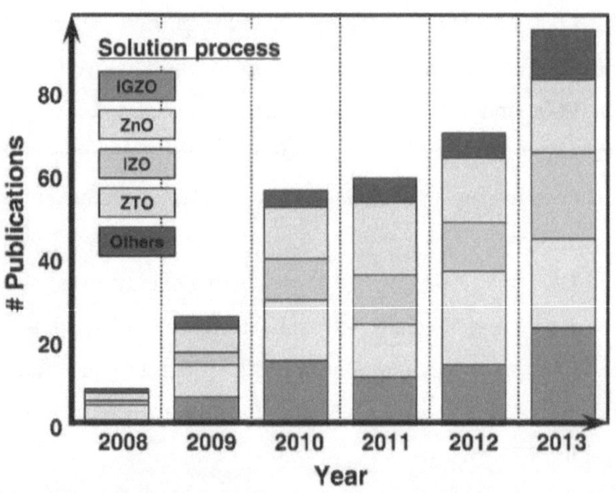

Figure 5 Percentage of publication amount in dependency of the investigated semiconductor material in the years 2008-2013. Adapted with permission[93]. Copyright (2014) The Japan Society of Applied Physics.

The following list summarizes the superior properties of multinary amorphous metal oxides in comparison to silicon as active channel material:

1) **Applicable with high-k dielectrics**: Various metal oxide high-k dielectrics like Al_2O_3, Y_2O_3, ZrO_2 and HfO_2 as well as their multinary hybrids can be utilized for the fabrication of AOS based TFTs due to the high conduction band (CB) offsets with numerous established metal oxide semiconductors. Thereby, the ionic nature as well as the amorphous state at low temperatures of the metal oxide dielectrics result in improved interactions with the semiconductor and thus exhibit fewer electronically defects at the dielectric/semiconductor interface. As a consequence, low electron trap densities and thus low leakage current densities can be realized.

2) **High electron mobility**: Silicon based TFTs exhibit field effect mobilities of $\mu_{FE} \approx 1$ cm^2 V^{-1} s^{-1}, while AOS based TFTs exhibit much higher mobilities of $\mu_{sFE} > 10$ cm^2 V^{-1} s^{-1}, which can be further increased by the modification of the materials composition and improvement of the processing conditions.[1]

3) **High current on/off ratio:** The use of AOS results in high on/off current ratios ($I_{on/off}$) which is in accord with the demand for higher refresh rates for next-generation displays. This is due to the fact that AOS are n-type semiconductors, thus a simple source-drain electrode deposition enables the lowering of the off-current and simultaneously raising the on-current. On the other hand, TFTs based on silicon as semiconductor require the generation of a p-n junction, whereby the source-drain electrodes suppress the TFT operation during the switching between the p- and n-type region.

4) **Low operation voltage:** The use of AOS enables the operation of the TFTs at voltages of ≤ 5 V, exhibiting a sub-threshold slope of ~ 0.1 V dec^{-1}.[94, 95] This is due to the amorphous state and the band-like charge transport of the AOS, leading to a reduction of the defect concentration within the band gap in comparison to silicon based semiconductors.

5) **Surface morphology:** AOS typically exhibit a very smooth surface roughness of $R_{RMS} < 0.5$ nm and low porosity which is very beneficial for the overall device performance. The amorphous state contributes to low leakage current densities due to the absence of grain boundaries. Furthermore, AOS can be homogenously deposited over large-area substrates.[96, 97]

6) **Low processing temperature:** The formation of functional AOS thin films is accessible at post-deposition annealing temperatures of $T \leq 350$ °C, depending on the chosen deposition technique. Thus, AOS are potential candidates as active channel material in TFTs for the fabrication of flexible electronics.[27, 98, 99]

3 High-*k* dielectrics

3.1 Fundamentals of high-*k* dielectrics

Dielectric materials are electrical insulators integrated in electronic circuits, e.g. as component in TFT devices. In contrast to conductors and semiconductors, electrical charges cannot move through a dielectric material, when it is located in an electric field, due to the large band gap energy (Figure 6), which is typically above 4 eV.

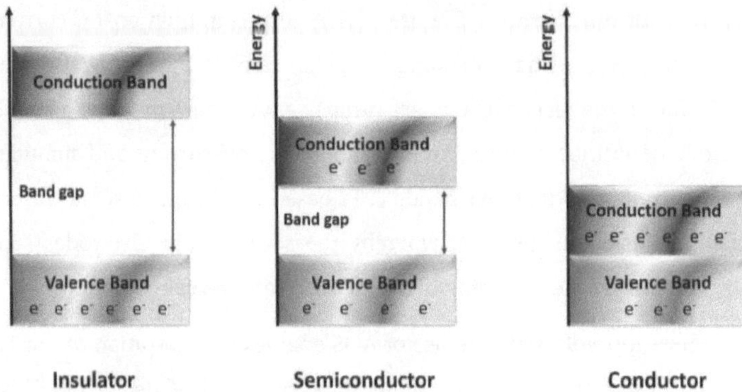

Figure 6 Schematic representation of the band gap energy of insulators, semiconductors and conductors.

The response of a dielectric material towards an applied electric field can be illustrated by the simple model of a parallel-plate or metal-insulator-metal (MIM) capacitor, with a dielectric material functioning as electrically insulating material between two metallic plates (Figure 7). As displayed in Figure 7c (top), without an external bias the corresponding charges are randomly distributed, showing no directional effect. Once a bias is applied, e.g. a positive potential to the top electrode, the dielectric material gets polarized. Thereby, the electric charges slightly shift from their equilibrium position, whereby positive charges shift in direction of the electric field and negative charges shift in the opposite direction. As a result, a net dipole moment throughout the whole material is generated, increasing the surface charge density at both sides of the dielectric material (Figure 7c, bottom) and decreasing the strength of the external electric field.[17]

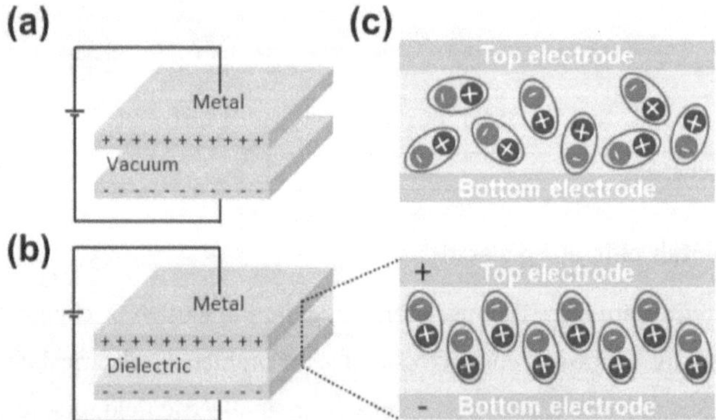

Figure 7 Schematic illustration of parallel-plate capacitors (a) without and (b) with dielectric layer. (c) Illustrated dipole polarization with (top) and without (bottom) an external electrical bias. Adapted with permission[17]. Copyright 2017, Elsevier B.V.

This phenomenon is generally known as dielectric polarization and is based on various mechanisms, including the electronic polarization, ionic polarization and orientation polarization (Figure 8).[17]

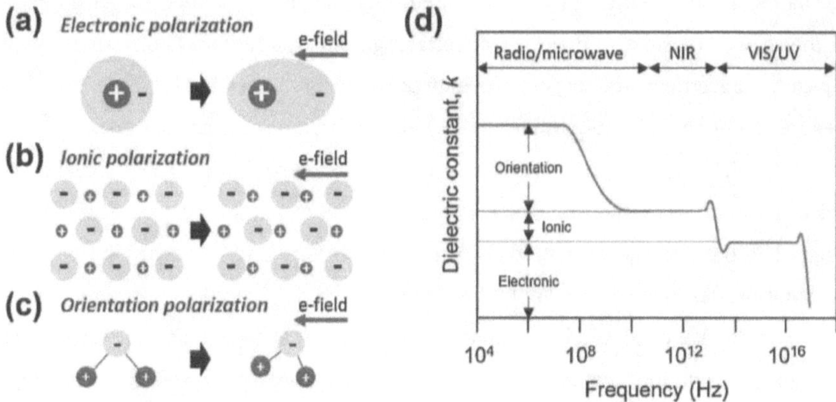

Figure 8 Schematic illustration of the dielectric polarization mechanisms: (a) electronic polarization, (b) ionic polarization, (c) orientation polarization. (d) Contribution of electronic, ionic, and orientation polarization regarding the dielectric constant at various frequencies. Adapted with permission[17]. Copyright 2017, Elsevier B.V.

The various polarization mechanisms are depending on the nature of the involved polar components within the dielectric material, which are rearranged under an external electric field, in dependency of the electric field strength. [17, 100]

Regarding the electronic polarization mechanism, the materials are composed of positively charged nuclei and negatively charged electron-clouds (Figure 8a). Thereby, the application of an external electric field results in a distortion of the symmetrical orbital paths, leading to a separation of positive and negative charges, which induces the dipole formation.

In case of the ionic polarization the mechanism is characterized by an induced net dipole moment, due to the displacement of cations and anions under an applied electric field (Figure 8b). In contrast, the orientation polarization occurs by the alignment of permanent dipoles along the direction of the external electric field (Figure 8c).

The contribution of the various polarization mechanisms are additive, but the individual mechanisms respond differently with respect to the applied frequency of the varying electric field (Figure 8d).[17] Regarding conventional TFT operations, the relevant frequency range is typically between $1\text{-}10^6$ Hz, in which the orientation polarization and electronic polarization are the dominant mechanisms. If the frequency exceeds 10^{10} Hz, the orientation polarization process does not contribute to the overall dielectric polarization anymore, while the electronic polarization even responds to very high frequencies up to 10^{16} Hz (Figure 8d). The dipoles of the gate dielectric in TFT devices must be able to reorient rapidly while employing frequently alternating electric fields. Thus, the required frequency region should be considered carefully for the choice of a suitable dielectric material regarding the respective electronic application.[17] Typically, dielectric materials possess a high polarizability, whereby the amount of this property is expressed by the relative permittivity (ε_r) also called dielectric constant (k).

The term "high-k dielectrics" basically refers to materials with a dielectric constant above 3.9, which is the dielectric constant of SiO_2. Another key performance parameter of a dielectric material is the leakage current density (J), which is mainly governed by two conduction mechanisms, described by the Poole-Frenkel effect and the Schottky emission.[63, 101-103]

The Poole-Frenkel effect is a bulk-limited theoretical model describing the electron transport in electrically insulating materials under a large electrical field, which is relying on the contribution of a certain amount of electrical traps within the bulk of the material (Figure 9).

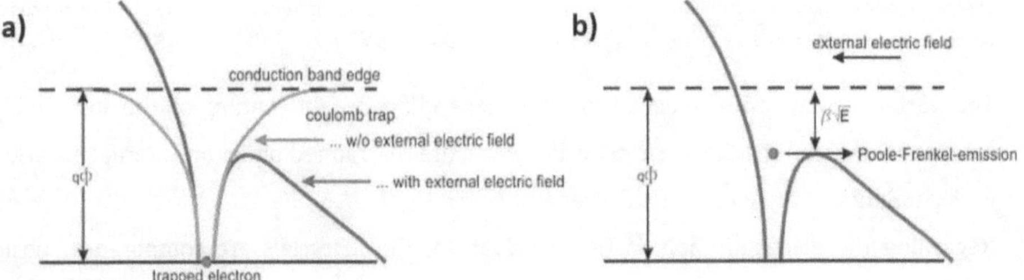

Figure 9 Schematic illustration of the Poole-Frenkel effect, a) change of the trap geometry at the transition from equilibrium to the interaction with an external electric field, b) reduction of the potential barrier caused by the external electric field, which increases the probability of the electron to promote from the trapped state to the conduction band. Adapted with permission[104]. Copyright 2013, AIP Publishing.

Generally, electrons are trapped in localized states e.g. in a single atom, unless randomly occurring thermal fluctuations supply sufficient energy to promote the electron from its localized state to the conduction band, followed by relaxation into a different localized state. Thereby, the electrons require less thermal energy to be promoted into the conduction band, when a large electric field is applied, due to a reduction of the Coulomb potential barrier within the bulk. As a consequence, electrons will move more frequently to the conduction band at higher electric fields, exhibiting an exponential increase of the leakage current with the square root of the applied voltage.[101]

The Schottky emission, also known as thermionic emission, is the governing mechanism at low electric fields and can be described by a similar theoretical model as the Poole-Frenkel effect, but refers to lowering the energy barrier between the metal electrodes and the dielectric material, which is caused by the electrostatic interactions with the electric field at the metal-insulator interface [63, 103] (Figure 10). Thus the leakage current, generated due to the Schottky emission is an electrode-limited process, which is highly dependent on the barrier height between the dielectric and the electrode.[103] Therefore, the Schottky emission represents a conduction mechanism in dielectric materials due to the reduction of the work function by an increasing electric field or temperature and thus contributing to the leakage current.

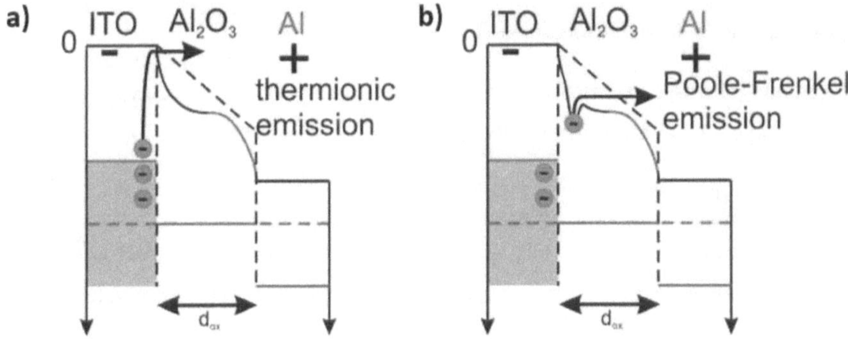

Figure 10 Schematic illustration of the mechanism of a) the Schottky emission in comparison to b) the Poole-Frenkel emission for an Al₂O₃ dielectric, using ITO as gate-electrode and Al as top-electrode. Adapted with permission[104]. Copyright 2013, AIP Publishing.

3.2 Capacitors

In order to evaluate the quality of such dielectric materials, capacitor devices were fabricated, allowing to determine the dielectric properties with respect to the capacitance per unit area (C_i), leakage current density (J) and breakdown voltage (E_B), which represent the most meaningful characteristic parameters regarding dielectric materials.

A capacitor is described as an incremental passive electronic device with two terminals (Figure 11). In principle, capacitor devices possess the ability to store electrical energy when an electric field is applied in a direct current (DC) circuit, but it is not possible to gain power or amplify signals, by the employment of capacitors.

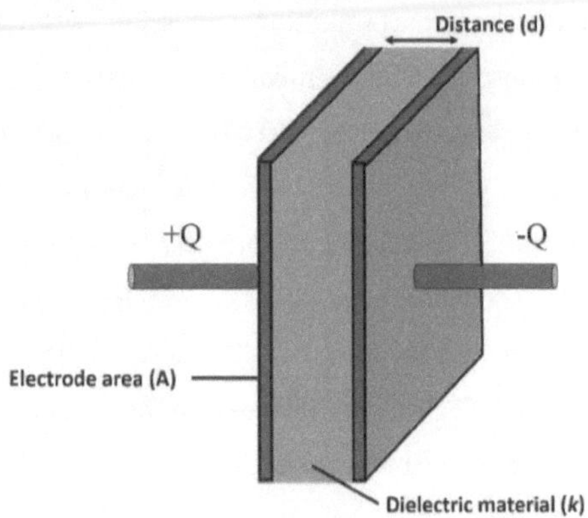

Figure 11 Schematic illustration of a parallel-plate capacitor.

The amount of stored electrical charge (Q), in dependency of the applied voltage (V) is defined as capacitance (C) and can be described as:

$$C = \frac{Q}{V} \qquad [F = \frac{As}{V}]$$

Equation 3.1

In contrast, when the capacitor is applied in an alternating current (AC) circuit, the device functions as an AC resistor with a frequency (ω) dependent impedance value (Z). It directs alternating voltages and alternating currents (AC) with a shift in the phase relationship between voltage and current, whereby the current leads the voltage by 90°, due to the charge storage capacity. The capacitance can be formulated according to the following equation.[105]

$$C = \sqrt{\frac{\frac{1}{|Z|^2}}{\omega^2 \cdot \frac{1}{tan^2(\theta)}}}$$

Equation 3.2

The capacitance can also be described by equation 3.3, using the model of a parallel plate capacitor, containing the bottom and top electrodes (A), which are separated by the dielectric layer (D). The capacitance is usually expressed as per unit area (C_i), which can simply be obtained by division of the capacitance (C) with the employed electrode area (A), in order to ensure comparability of the capacitance values (Equation 3.4). C_i is typically expressed in the unit [nF/cm²] for metal oxide thin film based capacitors.

$$C = \frac{\varepsilon_0 \cdot \varepsilon_r \cdot A}{D}$$

Equation 3.3

$$C_i = \frac{\varepsilon_0 \cdot k \cdot}{D}$$

Equation 3.4

C = capacitance [F]

A = electrode area [m²]

D = thickness of dielectric [m]

ε_0 = vacuum permittivity [/]

ε_r = relative permittivity [/] or dielectric constant (k)

The dielectric constant (k) is a material property, describing the permittivity of the dielectric in comparison to the vacuum permittivity (ε_0) and is therefore also called relative permittivity (ε_r). The dielectric constant is a complex function, depending on parameters such as temperature, strength of the electric field as well as the frequency and is directly related to the amount of capacitance (see Equation 3.4). It is evident that thinner dielectric films (D), for a given dielectric constant (k), result in increasing capacities per unit area (C_i). Ideal dielectrics are characterized by a constant capacitive behavior, regardless of the applied frequency, but in practice, dielectrics rather tend to possess higher capacities at low frequencies, due to the capacitive reactance X_C. The capacitive reactance describes the imaginary part of the complex impedance of the capacitor which is in dependency of the frequency and provided by the following equations.

$$Z = R + X_C \qquad \text{Equation 3.5}$$

$$X_C = 1/(2\pi \cdot f \cdot C) \qquad \text{Equation 3.6}$$

Figure 12a illustrates the comparison of dielectrics with a strong and weak frequency dispersion. Besides, dielectric properties such as the leakage current density (J) and breakdown voltage (E_B) are the key parameters, both can be extracted from I-V measurements. The leakage current density describes the amount of current which diffuses through the bottom gate electrode instead of contributing to the generated electric field. As a result, a higher power consumption is required deteriorating the overall device performance. Thus it is of highest importance to minimize the leakage current in order to achieve high performance and economically efficient electrical devices.

The breakdown voltage (E_B) describes the amount of voltage, dependent on the dielectric layer thickness, at which the material becomes conductive and loses its insulating properties. Thus, the breakdown voltage should be reasonable high, in order to ensure a certain stability of the final device. Figure 12b depicts the I-V characteristics of various dielectrics exhibiting the leakage current behavior as well as the electrical breakdown fields.

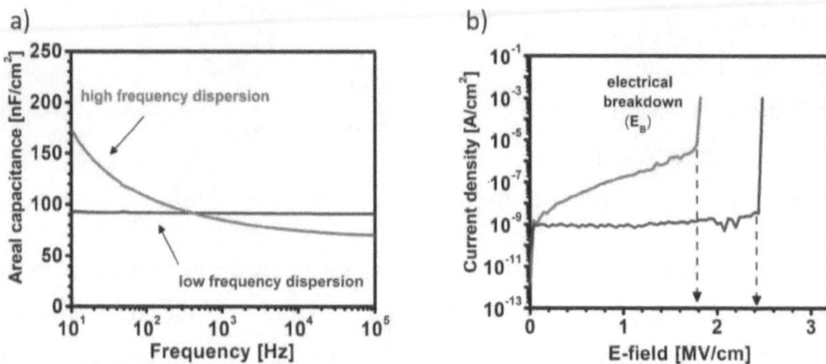

Figure 12 a) Representative capacity vs. frequency curves of solution-processed dielectrics implemented in capacitor devices, b) I-V characteristics with representative current leakage density vs. frequency curves and indicated breakdown fields.

In fact, the atmosphere in which the measurements are performed is crucial for the evaluation of the obtained dielectric properties. This is due to water molecules from the ambient atmosphere which adsorb on the thin films and contribute to a parasitic resistance, resulting in a remarkably increased areal capacitance. This effect becomes stronger at lower annealing temperatures, because more detrimental residual water and hydroxyl groups are residing on the surface of the metal oxide thin film. This enhances the adsorption of additional water molecules. On the other hand, the leakage current and electrical breakdown properties will be degraded, due to water adsorption. As a consequence, it is difficult to compare the dielectric properties measured in an ambient atmosphere with those measured under an inert atmosphere. Thus, in order to evaluate the properties of the pristine metal oxide dielectric, all samples within this book were measured under an inert atmosphere in a glove-box (O_2, H_2O < 0.5 ppm) to ensure comparable surface conditions.

3.3 Thin-film transistors (TFTs)

A thin-film transistor is a three-terminal device, which is operating under an electric-field and thus is categorized as field-effect transistor (FET). TFTs are usually employed as switch in digital logic circuits and basically consist of a gate electrode, the source and the drain electrodes (the three terminals) as well as a dielectric layer and active semiconductor channel material. These components can be arranged in different geometries, with the gate and the source/drain electrode either at the top or bottom of the stack (Figure 13), whereby each of these geometries possess particular advantages and disadvantages, depending on the utilized materials.[106]

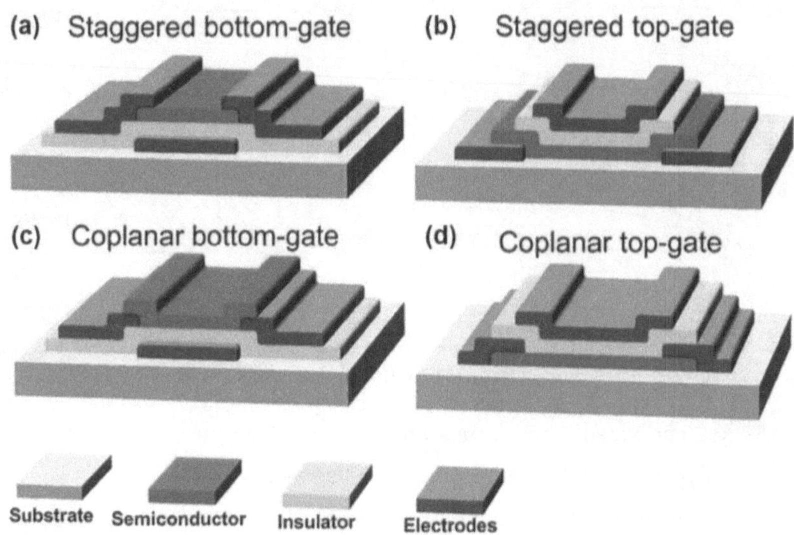

Figure 13 TFT device geometries: (a) staggered, bottom-gate top-contact (BGTC), (b) staggered, top-gate bottom-contact (TGBC), (c) coplanar, bottom-gate bottom-contact (BGBC) and, (d) coplanar, top-gate top-contact (TGTC). Adapted with permission[106] Copyright 2012, Wiley-VCH Verlag GmbH & Co. KGaA, Weinheim.

The functionality of the device is based on the ability to control the current flow through the semiconductor layer between the source and the drain electrodes, induced by the generated electric-field between the gate and drain electrodes.

It is of crucial importance to ensure that no current flows between the gate and the drain electrode, thus the incorporation of a dielectric layer between these electrodes is required. The dielectric layer should provide excellent insulating properties such as a low gate leakage current density, while possessing a high dielectric constant. High dielectric constants contribute to a high capacitance per unit area as well as to a reduced amount of voltage in order to operate the TFT. The choice of the high-k dielectric plays a key role since the flow of charge carriers in TFTs is based on the accumulation of charges within the semiconductor layer, generated by the electric field at the dielectric/semiconductor interface.

In order to evaluate the quality of the thin-film transistors, TFT performance characteristics extracted from the transfer and output curves (I-V characteristics) are determined. Representative examples of the transfer and output characteristics are shown in Figure 14.

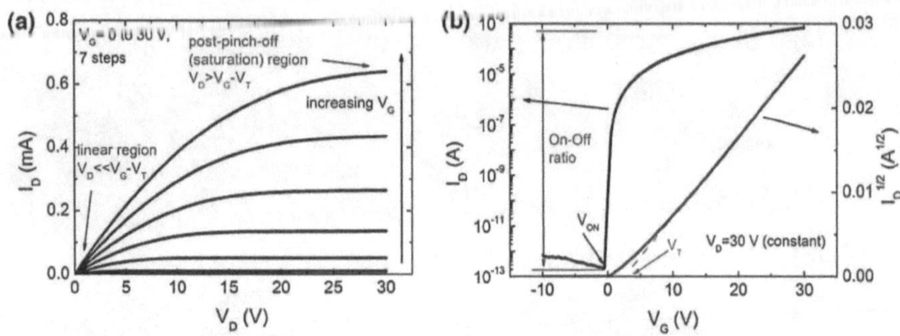

Figure 14 Electrical characterization of a thin-film transistor (a) output curves (I_D against V_D) and (b) transfer curve (I_D against V_G). Adapted with permission[106]. Copyright 2012, Wiley-VCH Verlag GmbH & Co. KGaA, Weinheim.

The TFT key performance parameters are represented by the on-off current ratio ($I_{on/off}$), field-effect mobility (μ_{FE}), threshold voltage (V_{th}) as well as on-voltage (V_{on}) and subthreshold swing (SS). The on-off current ratio ($I_{on/off}$) is simply defined by the ratio of the current in the on- and off-state of the transistor device. The field-effect mobility (μ_{FE}) of a thin-film transistor is directly related to the amount of charge carriers available within the active semiconductor material.[106] Besides charge carriers intrinsically present within the semiconductor material, the transport mechanism is also determined by scattering of charge carriers due to e.g. grain boundaries, point defects and ionic impurities.[107] Regarding thin-film transistors, the charge carrier mobility is calculated depending on the voltage applied across the source-drain (V_{DS}) electrodes as well as the gate-source electrodes (V_{GS}). The field-effect mobility can be extracted either at low voltages, in the linear regime or at higher voltages in the saturation regime, which are therefore called as linear field-effect mobility (μ_{lin}) and saturation field-effect mobility (μ_{sat}), respectively. The calculations for both cases are given by the following equations.[108, 109]

- linear field-effect mobility (μ_{lin}): at low V_{DS} ($V_{DS} \sim 0 \ll V_{GS}$)

$$I_{DS} = \frac{W}{L} \mu_{lin}.C_i \left[(V_{GS} - V_{th}).V_{DS} - \frac{V_{DS}^2}{2} \right]$$

Equation 3.7

- saturation field-effect mobility (μ_{sat}): at high V_{DS} ($V_{DS} \gg V_{GS} - V_{th}$)

$$I_{DS} = \frac{W}{L} \mu_{sat}.C_i (V_{GS} - V_{th})^2$$

Equation 3.8

W = channel width

L = channel length

C_i = capacitance per unit area (dielectric layer)

V_{DS} = drain-source voltage

V_{GS} = gate-source voltage

V_{th} = threshold voltage

The threshold voltage (V_{th}) is defined as gate voltage at which a significant current, formed at the dielectric/semiconductor interface, flows between the source and the drain electrodes through the active semiconductor channel material. The V_{th} value can be extracted from the $I_{DS}^{1/2}$ vs. V_{GS} plot, at the intersection with the x-axis (V_{GS}). In contrast, the on-voltage (V_{on}) is described as the gate voltage at which the current starts to flow through the active channel material and turn-on the device. The subthreshold swing (SS) is defined as the inverse of the maximum slope of the transfer curve. The amount of SS describes the rate at which I_{DS} increases by one decade while the applied voltage V_{GS} constantly increases. Thereby, small values of SS (usually > 0.5 V dec^{-1}) are feasible for low power consumption microelectronic devices.[106] The subthreshold swing is given by Equation 3.9.[110-112]

$$SS = \left(\frac{d \log I_{DS}}{d V_{GS}} \right)^{-1} \qquad\qquad \text{Equation 3.9}$$

3.4 The choice of metal oxide high-k dielectrics

As discussed in section 1.2 high-k dielectrics based on metal oxides are potential candidates as the gate dielectric in oxide based thin-film transistor (TFT) devices and are close to replace the conventional SiO_2 dielectric.[42, 113]

In order to determine the most suitable high-k dielectric, we can choose in principle from a plethora of possible materials across the periodic table. The following chart (Figure 15) illustrates the selectable binary oxide materials, which possess higher dielectric constants than SiO_2.[114]

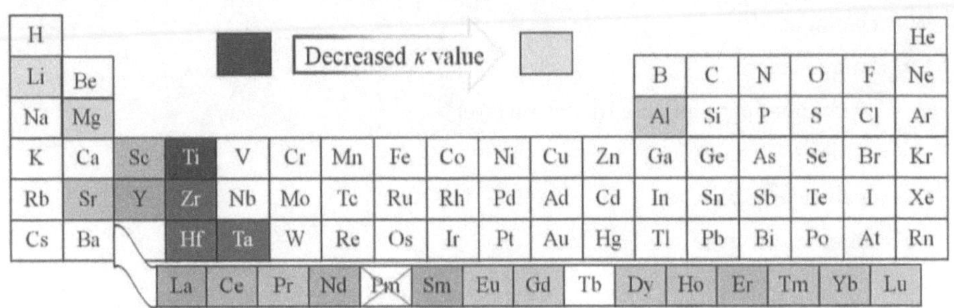

Figure 15 Periodic table of the elements. The binary oxides with higher k values than SiO₂ are highlighted; the color scale indicates the amount of the dielectric constant.. Adapted with permission[114]. Copyright 2018, Wiley-VCH Verlag GmbH & Co. KGaA, Weinheim.

Thereby, the dielectric must possess specific material properties in order to provide excellent dielectric characteristics and ensure feasibility regarding the integration within existing industrial process circumstances. The following list summarizes the necessary requirements regarding the choice of a suitable metal oxide gate dielectric for replacing SiO_2, which is thoroughly helpful to restrict the potential candidates. [34, 36, 42] Details regarding these requirements will be discussed subsequently.

1) The material is deposited onto the Si substrate, so it must be thermodynamically stable with it.
2) The material must possess a dielectric constant (k ≥ 10) which is high enough to ensure high capacities and further down-scaling, in accord with economical demands.
3) The material must possess a band gap that is high enough (> 5 eV) to ensure low current leakage densities ($J < 1 \times 10^{-8}$ A cm⁻²), large breakdown fields ($E_B > \sim 4$ MV cm⁻¹) and high CB offsets with the active semiconductor material (> 1 eV).
4) The material must maintain the amorphous phase over a broad temperature range.
5) The material must provide a low surface roughness.
6) The material must exhibit a low concentration of electronically active defects.

Once SiO_2 is replaced as gate dielectric in TFT devices, Si loses its main advantage as active semiconductor material, but still remains the substrate, also functioning as gate material for technologically relevant fabrication processes, due to its superior mechanical properties and abundance. Anyways, the industry is already committed to large-area silicon wafers keeping it a relevant material for integrated circuits. Thus the interface properties between the high-k oxide and Si still must be considered, in order to achieve high performance TFTs.[42]

Since the high-k oxide is deposited onto the Si substrate, it must be thermodynamically stable with it, preventing the generation of SiO_2 or silicides, described by the following reactions:

$$MO_2 + Si \rightarrow M + SiO_2$$
$$MO_2 + 2\,Si \rightarrow MSi + SiO_2$$

The generated SiO_2 layer would increase the equivalent oxide thickness (EOT) and thus counteracts the desired effect of the high-k oxide, accompanied with an uncertain reproducibility. The generation of silicides would have even more severe consequences, since they could short circuit the device, due to their often metallic nature. As a result, the respective high-k oxide must exhibit a greater heat of formation than SiO_2.[42] Thus, regarding the deposition on Si substrates, potential candidates such as Ta_2O_5, TiO_2 as well as the ternary oxides $BaTiO_3$ and $SrTiO_3$ are ruled out right from the start, due to their thermodynamic instability with silicon.

According to Schlom and Haeni[115], this criterion alone limits the reasonable choices to the metal oxides CaO, SrO, BaO, Al_2O_3, ZrO_2, HfO_2, Y_2O_3, La_2O_3 and the lanthanides such as Pr_2O_3[116], Gd_2O_3[117] and Lu_2O_3[118] which are chemically very similar to La_2O_3, without possessing particular advantages. Unfortunately, the group (II) elements SrO, CaO, BaO are extremely reactive with water, which renders them less suitable as gate oxide material in such devices, leaving us with the binary oxides Al_2O_3, ZrO_2, HfO_2, Y_2O_3 and La_2O_3 as potential gate dielectrics in order to replace SiO_2. Ultimately, the dielectric material must provide excellent dielectric characteristics such as a high capacitance per unit area (C_i) and thus the fundamental requirement for the dielectric oxide material is a sufficiently high dielectric constant which should be at least $k \geq 10$. As a result, thicker dielectric layers in comparison to SiO_2 can be employed, while maintaining an equal or even higher areal capacitance[119] and additionally ensuring the possibility of down-scaling, inevitable for future performance enhancements.

Another essential requirement represents the band gap energy (E_g) of the metal oxide, which should be as high as possible, because it directly determines the leakage current density as well as the CB offset with the semiconductors. Furthermore, the magnitude of the band gap energy is also related to the electrical breakdown field.[119] Typically, there is an inverse trend between the value of the dielectric constant and the amount of the band gap energy. Materials with high k values tend to possess smaller band gaps and vice versa (Figure 16), which is the reason why we have to accept a rather low k value for a binary metal oxide in order to ensure sufficiently high band gaps and thus a low leakage current density, large breakdown field and high CB offset with the semiconductor.[120]

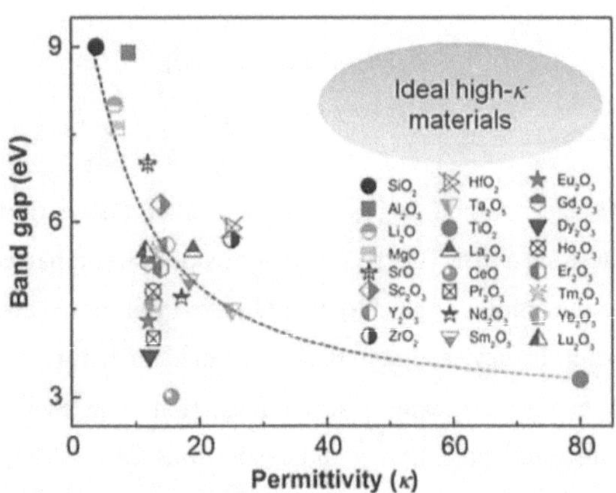

Figure 16 Dielectric constant (k) and band gap energy (E_G) of various binary oxide dielectrics. Adapted with permission[114]. Copyright 2018, Wiley-VCH Verlag GmbH & Co. KGaA, Weinheim.

According to Wager *et al.* an ideal dielectric material should possess a current leakage density of $J < 1 \times 10^{-8}$ and a breakdown field of $E_B > \sim 4$ MV cm^{-1}, even when down-scaling to very thin dielectric layers.[119]

The high CB offset with the semiconductor ensures that the dielectric material acts as an insulator, since the potential barrier at each band must be over 1 eV in order to minimise carrier injection by the Schottky emission of electrons or holes into the oxide bands.[120] As a result, only high-k oxides with band gaps over 5 eV fulfil this requirement, regarding the CB offset with commonly used AOS. Taking these criteria into account the potential binary metal oxides still include ZrO_2, HfO_2, Al_2O_3, Y_2O_3 and La_2O_3.

The dielectric material should be able to maintain the amorphous phase over a wide temperature range (T $\leq$ 500 °C), because crystalline grain boundaries can act as preferential pathways for impurity diffusion, thus contributing to the leakage current density. In general, the use of amorphous metal oxides (AMOs) is advantageous in many ways. The amorphous phase promotes a smooth surface of the metal oxide thin films, which can prevent surface scattering.[36] Furthermore, amorphous thin films which possess smooth surfaces are highly desirable regarding TFT applications due to an improved dielectric/semiconductor interface, which results in lower leakage current densities as well as a higher degree of reproducibility.[67, 121] Additionally, AMOs are able to configure the interface bonding to minimize the number of interface defects. Furthermore, the dielectric constant of an AMO is isotropic, which is preventing the carriers to scatter due to fluctuations in polarization from differently oriented oxide grains. Besides, AMOs allow to vary the composition by maintaining the phase, e.g. in alloys, interfacial layers, or concerning dopants and require lower processing temperatures, which is contributing to low manufacturing costs.[36]

Regarding the potential candidates, the binary metal oxides Y_2O_3, HfO_2 and ZrO_2 tend to crystallise at lower temperatures (~400 °C) often resulting in a nano-crystalline phase formation. Additionally, HfO_2, ZrO_2 and especially La_2O_3 are very hygroscopic and thus adsorbing higher amounts of water, which contributes in a negative manner to the overall device performance, due to an enhanced metal hydroxide formation.[122]

As a result, Al_2O_3 remains in the list and might be the most promising binary metal oxide as gate dielectric for thin-film transistor applications, although it possesses the lowest dielectric constant ($k \approx 10$) among the remaining candidates.

In 2014, Wang *et al.* performed a systematic investigation of In_2O_3 and InZnO based TFTs, with varying solution-processed gate dielectrics, employing Al_2O_3, ZrO_2, Y_2O_3 and TiO_2, which have been deposited via spin-coating and processed at 300 °C. Thereby, the utilization of the Al_2O_3 dielectric could achieve the best overall TFT performance. According to Wang *et al.* the superior electrical performance can be attributed to a minor accumulation of charge carriers within the bulk of the dielectric as well as a small trap density at the dielectric/semiconductor interface.[63]

Consequently Liu *et al.* demonstrated in the year 2015, the synthesis of high performing Al_xO_y thin films, derived from an aluminum nitrate nonahydrate precursor. The employment of a 15 nm thick Al_xO_y film, processed at a post-deposition annealing (PDA) temperature of 350 °C, is exhibiting a very low leakage current density of $J < 1 \times 10^{-9}$ A cm^{-2} and a satisfying electrical breakdown field of $E_B = 5.5$ MV cm^{-1}. Furthermore, the solution-processsed Al_xO_y dielectric possess a very high areal capacitance of 413 nF cm^2, by exhibiting a dielectric constant of $k = 7.0$.[14]

Nevertheless, also the other potential candidates regarding the gate dielectric such as ZrO_2, HfO_2 and Y_2O_3 showed good results with respect to the dielectric properties. Song *et al.* fabricated solution-processed Y_xO_y dielectrics by using yttrium nitrate hydrate as precursor. In this work, reasonable dielectric characteristics could be achieved at an annealing temperature of 300 °C. The capacitance per unit area amounts 71.7 nF cm^{-2} (d= 188 nm), while exhibiting a leakage current density of $J = 5.2 \times 10^{-8}$ A cm^{-2}. Interestingly, the employment of higher processing temperatures results in higher leakage current densities, due to an increasing degree of crystallinity of the Y_xO_y dielectric and thus deteriorating the I-V characteristics.[9]

Table 1 shows an overview of the electrical properties of some selected binary oxides for the application as gate dielectric in TFT devices.

Table 1: Overview of electrical properties of selected binary metal oxide dielectrics.[32, 36, 123, 124]

Material	Dielectric constant (k)	Band gap energy [eV]	CB offset Si [eV]	CB offset ZnO [eV]	CB offset In$_2$O$_3$ [eV]	CB offset SnO$_2$ [eV]
SiO$_2$	3.9	9.0	3.2	4.3	4.6	4.8
Al$_2$O$_3$	9	8.8	2.8	3.6	2.9	3.1
Sc$_2$O$_3$	13	6.0	2.3	3.1	3.7	3.9
Y$_2$O$_3$	15	6.0	2.3	3.1	3.7	3.9
La$_2$O$_3$	30	6.0	2.3	3.1	3.7	3.9
HfO$_2$	25	5.8	1.5	2.1	2.4	2.6
ZrO$_2$	24	5.8	1.4	2.0	2.3	2.5

3.5 Multinary metal oxide high-*k* dielectrics

The binary metal oxides HfO$_2$[4, 78, 125-127], ZrO$_2$[5-7, 63, 78, 128], Y$_2$O$_3$[8-12, 63, 129] and Al$_2$O$_3$[2, 12-15, 18, 63, 130-138] are promising materials for the future application as gate dielectrics in TFT devices and have already demonstrated very good results with respect to their dielectric properties in capacitor and TFT devices, respectively. As discussed in the previous section, all binary oxide dielectrics suffer from a significant drawback. Binary oxide dielectrics with higher dielectric constants usually possess narrow band gaps and thus display higher leakage currents, deteriorating the overall device performance, due to an inverse trend between the dielectric constant and the band gap energy. A possible strategy to solve this problem is the formation of hybrid multinary metal oxide dielectrics, e.g. ternary metal oxide dielectrics. The idea is to combine one dielectric material which possess a high dielectric constant with another material providing a wide band gap, resulting in a decreased leakage current density and consequently exhibiting superior electrical properties compared to the individual binary metal oxides. Furthermore, the generation of such hybrid metal oxide dielectrics allows to fine-tune the electrical properties by a compositional variation of the respective metal oxide components. Regarding the wide band gap material SiO_2 (E_g = 9.0 eV) and Al_2O_3 (E_g = 8.8 eV) are very promising. Regarding the material possessing a high dielectric constant, metal oxides like HfO_2 ($k \approx 25$), ZrO_2 ($k \approx 25$), La_2O_3 ($k \approx 30$) and Y_2O_3 ($k \approx 15$) are favorable. Thus, various material combinations for a high performance ternary metal oxide dielectric such as Hf_xSi_yO[139], Hf_xAl_yO[140-143], Zr_xAl_yO[144-147], La_xAl_yO[148] and Y_xAl_yO[149-151] are currently subject of intensive research. When comparing the wide band gap oxides SiO_2 and Al_2O_3, aluminum oxide might be the better choice due to its higher dielectric constant, comparable band gap energy and sufficiently high CB offsets with common semiconductors (see Table 1), while SiO_2 might decrease the overall dielectric constant of the hybrid material too drastically. Among the high-k components HfO_2, ZrO_2 and especially La_2O_3 are very hygroscopic, resulting in an increased transformation of the respective oxide into the hydroxide, deteriorating the final device performance. Yttrium oxide, Y_2O_3 on the other hand represents an interesting counterpart for the Al_2O_3 based hybrid material due to a combination of desirable electrical properties like a high dielectric constant (14–18), high breakdown voltage (>3 MVcm^{-1}), high refractive index (1.9–2.0) and low dissipation factor (<0.005), while possessing a reasonable high band gap energy (5.8 eV).[152] Furthermore, Y_2O_3 can form an intimate chemical surface bond with oxide semiconductors, resulting in an improved dielectric/semiconductor interface formation and thus enabling the realization of low electron trap densities and leakage currents.[153] Therefore, yttrium aluminium oxide (Y_xAl_yO) represents a very promising ternary hybrid material for high performance gate dielectrics in the realm of TFT devices. Therein, Y_2O_3 represents the high-k material (k $\approx$ 15)[152] and Al_2O_3 the component with the wide band gap (E_g = 8.8 eV).[154] Note, although Y_2O_3 tends to crystallize at T < 400 °C, the YAl_xO_y hybrid material will be amorphous in practice, up to a certain aluminum oxide amount within the composition.

Consequently, in the last couple of years research in the field of hybrid multinary metal oxide dielectrics gained attention, due to the tunable electrical properties of such materials. Jeong *et al.* reported in the year 2018 on solution-processed yttrium aluminium oxide thin films processed at an annealing temperature of 400 °C, exhibiting excellent dielectric properties.[149] Thereby, three different compositions of YAl_xO_y dielectrics were studied systematically and compared to the respective binary oxides of aluminum and yttrium. As a result and as expected, all YAl_xO_y compositions possessed higher dielectric constants compared to the pristine binary Al_xO_y dielectric. Additionally, all YAl_xO_y compositions exhibited higher electrical breakdown fields and interestingly lower leakage current densities, compared to both the pristine Y_xO_y and

Al_xO_y dielectric. The authors attribute the decrease of the leakage current density in such hybrid materials to the reduction of hydroxyl species within the YAl_xO_y layers, which corresponds to the amount of yttrium incorporation into the Al_xO_y lattice.[149]

These findings further consolidate the potential of hybrid multinary metal oxides for the future optimization of gate dielectrics in e.g. TFT devices.

4   Solution-processing of high-*k* dielectric thin films

The deposition procedure of functional metal oxide thin films such as dielectrics or semiconductors is generally classified in vacuum-based and solution-based deposition techniques.

The vacuum-based deposition techniques include atomic layer deposition (ALD)[43-48], pulsed laser deposition (PLD)[155-159], chemical vapor deposition (CVD)[49-52] as well as various sputter deposition techniques such as DC reactive magnetron sputtering[53-56] and radio frequency (RF) magnetron sputtering[57-59]. These vacuum-based deposition techniques enable the formation of high quality metal oxide thin films as well as a high degree of reproducibility, which is extremely desirable for electronic applications. Unfortunately, these vacuum-based processes often require expensive setups and feature low material deposition rates, resulting in low throughputs and thus render large-area depositions rather difficult and demanding.[16]

In contrast, solution-processing represents an interesting, alternative deposition technique, which is cost-efficient and allows large-area depositions of the respective metal oxide thin films via various different approaches.[20, 74, 75] Furthermore, solution-processing enables a simple approach to tune the material composition of the final functional metal oxide by dissolving the different metal oxide precursors in the desired ratio, resulting in multinary metal oxide thin films with optimized electrical performance characteristics. This strategy is not easily applicable for vacuum-based deposition techniques, since metal oxide targets with a predefined composition are used, complicating the optimization by tuning the material composition.

The solution-based process for obtaining metal oxide thin films can generally be divided into different processing stages, as illustrated in Figure 17, whereby each step has to be considered carefully in order to achieve high quality thin films.

Thereby, the first obstacle is the choice of a suitable precursor, which is of crucial importance since the physical and chemical properties of the precursor-solvent system have a huge impact on the quality of the generated metal oxide thin films and thus the final device performance. The importance and influence of the utilized precursors for the formation of the desired high-k dielectric or semiconductor thin films will be discussed in detail in the following sub-section of this chapter (section 4.1).

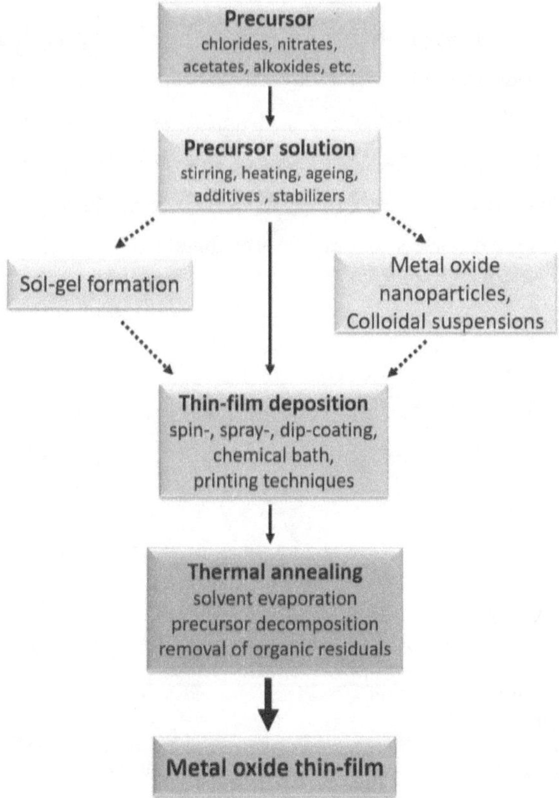

Figure 17 Schematic illustration of individual process steps involved in the solution-processing of metal oxide thin films.[160]

Subsequently, the generation of a stable precursor solution is required, which is compatible with the chosen deposition process and substrate material. For the generation of suitable precursor solutions, it is extremely beneficial to utilize precursor structures which ensure a high solubility in the desired solvent. Additionally, a certain long-time stability of the solution is of great importance and therefore the addition of another solvent and stabilizers as additives is often performed.[18, 63, 103, 130-133] Thereby, solution-processing enables the employment of a wide variety of precursors, additives and stabilizers. In case of alkoxide precursors, the sol-gel chemistry enables the transformation of the precursor/solvent system into a sol which gets further converted into a gel by forming a structural network of the sol.[161]

Alternatively, the precursor solution can be transformed into colloidal suspensions consisting of metal oxide nanoparticles, generated by e.g. microwave assisted synthesis, prior to the thin-film deposition.[16, 17, 160]

The deposition step itself plays another key role, in order to obtain homogenous, dense and crack-free metal oxide thin films, whereby the most commonly used approaches are spin-coating[9, 11, 13, 18, 63-65], spray-coating[12, 162], chemical bath deposition (CBD)[60-62] and inkjet printing[67-71, 73]. A schematically illustration of several different solution-based deposition techniques is shown in Figure 18. Finally, after the successful deposition, the homogenous precursor thin films get converted into the desired metal oxide thin films by thermal annealing.

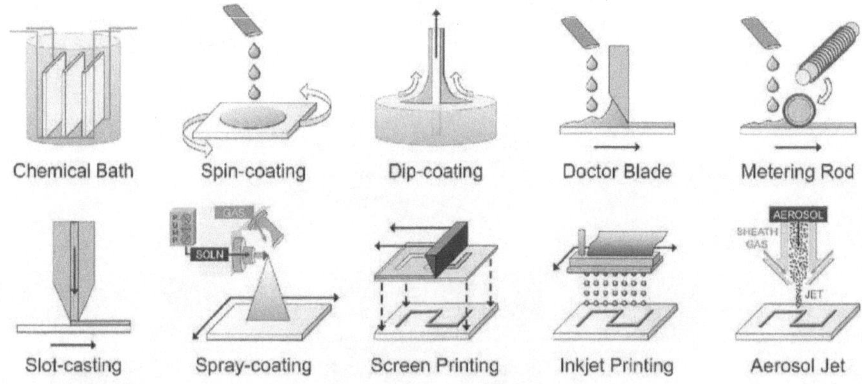

Figure 18 Schematic illustration of various solution-based deposition techniques. Adapted with permission[160]. Copyright 2011, Royal Society of Chemistry.

4.1 The role of the precursors for solution-processing of metal oxide thin films

The solution-processing of functional metal oxide thin films is primarily based on the deposition of stable precursor solutions, by employing various precursor systems. Thereby, the use of metal salts as precursors has been extensively investigated and was successfully applied to generate metal oxides with a variety of electronic applications such as ZnO, In_2O_3, SnO_2, Ga_2O_3 and their multinary hybrid compositions as active channel semiconductors in TFT devices.[163-184] Recently, also binary metal oxide gate dielectrics like Al_2O_3[2, 13-15, 18, 103, 122, 129-138], Y_2O_3[8-11, 129], ZrO_2[5-7, 128] and HfO_2[4, 125-127] as well as their multinary hybrids e.g. YAl_xO_y[149-151], $ZrAl_xO_y$[144-147] and $HfAl_xO_y$[140-143] have been successfully processed via the solution-based route.

For the formation of such metal oxide thin films, mostly metal salts are employed as precursors. Among the metal salts, nitrates, chlorides, carboxylates and alkoxides are the most prominent choices, mainly due to their commercial availability and low-cost.

Metal alkoxides are formally described as $M^{n+}(^-OR)_n$ and can be synthesized for a variety of metal cations (M^{n+}) by substitution reactions with the desired alkoxide (RO^-), which is the conjugate base of the respective alcohol (ROH). In comparison to the before-mentioned metal salts, the utilization of metal alkoxides results in metal oxide thin films exhibiting the lowest amount of residual contaminations, at low processing temperatures. On the other hand, metal alkoxides possess a high tendency of hydrolysis reactions due to the nucleophilic nature, which results in the necessary requirement of an inert atmosphere for the deposition process, in order to avoid the enhanced formation of metal hydroxides, rendering the whole fabrication process of these thin films less cost-efficient. A possible strategy which allows an ambient atmosphere coating is the addition of a chelating agent such as acetylacetone which functions as a stabilizer.[160] Thereby, the chemical modification of the metal alkoxide precursor solution with chelating agents is leading to a decrease of the hydrolysis and condensation reaction rates, resulting in an enhanced densification of the metal oxide thin films.

Acetylacetone, formally described as 2, 4-pentanedione, belongs to the class of ß-diketones, which are organic molecules possessing two keto groups separated by a methylene group. These ß-diketones commit the so called keto-enol tautomerizaton, whereby a hydrogen atom internally migrate to the oxygen atoms of the carboxylate functionality, forming a C-C double bond within the structure. The conjugated anion, acetylacetonate (acac), is able to coordinate bidental to the metal cations and thus forming chelate complexes. Thereby, the stability of the precursor solution is depending on the binding strength between the metal cation and the chelate ligand, in comparison to the utilized alkoxide ligand.[185, 186] In fact, such *in situ* formed complexes, derived from the added stabilizers are found to be present within the final metal oxide thin film and thus leading to uncertain electrical device performances[72], accompanied with the possible corrosion of the substrate material.[187]

Another precursor type is represented by metal carboxylates, which are the corresponding metal salts of the respective carboxylic acid and thus are formally described as $M^{n+}(RCOO^-)_n$. Thereby, the organic substituent "R" can be purposely designed with respect to the desired chemical properties. The carboxylate anion provides an additional resonance stability, due to the delocalized negative charge between the two oxygen atoms of the carboxylate functionality and thus is able to bind to the cation as monodentate or polydentate ligand, leading to different chemical structures of the coordination compound. The simplest carboxylate is the formate ion (HCOO⁻), with "R" representing a hydrogen atom. The most commonly used carboxylate among the versatile class of carboxylate precursors are acetates (CH_3COO^-).[160] Consequently, such carboxylates with short alkyl chains are highly soluble in polar solvents like water or 2-methoxyethanol (2-ME), which are established solvents for the spin-coating technique. Besides, it is possible to tune the solubility of the precursor by increasing the length of the alkyl chain, also suitable for non-polar solvents.[8, 188] The solubility of the precursor in established solvents is a significant issue, since a sufficiently high concentration is necessary to ensure a high density of the final metal oxide thin film. If the precursor concentration is too low, the metal oxide thin film usually consists of an enhanced degree of porosity and cracks within the microstructure of the films caused by solvent evaporation and the decomposition of ligand components from the precursor.[189] Therefore, it is useful to fill the voids and cracks by performing repeated iterations of the coating process and thus increasing the density of the resulting thin film.[189]

A frequently used precursor type are metal chlorides ($M^{n+}Cl^-_n$), which are extensively investigated for the formation of functional metal oxide dielectrics such as Al_xO_y and showing good results with respect to the dielectric properties.[2, 13, 135, 145] In 2015 Tan *et al.* demonstrated Al_xO_y-based capacitors derived from the aluminum chloride precursor, exhibiting good electrical performance characteristics, displaying a very low leakage current density of 2.6×10^{-9} A cm^{-2} at 3 V.[135] However, a major drawback of this study is the extreme high annealing temperature of 550 °C, unsuitable for low-temperature solution processing of flexible electronics. Typically, the use of metal chlorides for the formation of metal oxide dielectrics at low processing temperatures results in relative high leakage currents as reported from Avis and Jang by fabricating Al_xO_y-based capacitors at 300 °C, exhibiting J = 6.3×10^{-5} A cm^{-2} at 1 MV cm^{-1}.[13] In fact, regarding the above mentioned metal salts, especially the use of chlorides can cause complications regarding the synthesis process. Metal chlorides typically require elevated temperatures in order to decompose completely, due to the strong ionic nature of the bonds between the metal cation and the chloride anions. Furthermore, the calcination of the precursor solution leads to the formation of acidic by-products like hydrochloric acid (HCl) vapours, which cause an increased degree of porosity for the metal oxide thin films, resulting in deteriorated interface properties. Additionally, residual Cl⁻ ion traces are detectable within the final ceramic, which are able to trap charges at the interface and thus might deteriorate the overall device performance.[190-192]

In 2013, Lee *et al.* investigated the influence of the employed precursors in solution-processing of IZO based TFTs, by varying the zinc oxide precursor using zinc chloride, zinc acetate as well as a 1:1 precursor mixture. In this study the amorphous IZO films were obtained at post-deposition annealing temperatures of 450 °C. Thereby, the utilization of only zinc chloride resulted in TFTs with an increased field effect mobility, but a drastically deteriorated overall device performance as well as a high electrical loss. The IZO-based TFT employing the 1:1 mixture of the zinc precursors exhibited slightly degraded characteristics to those of the zinc acetate based devices, but exhibited a much higher positive-stress stability than the other devices. Therefore, this study verified that providing a suitable amount of chlorine during annealing may improve the device stability, but an excess of chlorine leads to unfavorable results, deteriorating the overall device performance drastically.[193]

In comparison to chlorides, metal nitrates ($M^{n+}(NO_3^-)_n$) require much lower decomposition temperatures due to the weaker strength of the ionic bonds between the metal cation and the nitrate anions. Consequently, at processing temperatures T > 350 °C, the resulting thin films contain no residual nitrogen content, deriving from the nitrate ligand and thus representing an attractive alternative to metal chloride precursors.[15, 194] Additionally, Keszler *et al.* could demonstrate in a systematic study that the solution-based route utilizing the aluminum nitrate precursor processed at 500 °C, results in Al_xO_y films with comparable dielectric properties to those obtained from atomic layer deposition.[195] Consequently, the utilization of metal nitrates is widely used for the synthesis of various functional metal oxides demonstrating good electrical characteristics.[9, 10, 14, 15, 18, 63, 134, 196] These findings clearly proof the importance of the precursor choice for the synthesis of high quality metal oxide thin films via solution processing. However, intrinsic precursor properties, the requirement for stabilizers and other additives as well as high processing temperatures still restrict the industrial application regarding large-area fabrication.[18, 19]

A promising class of precursors are represented by single-source molecular precursor compounds, which undergo a clean decomposition under thermal treatment, by releasing volatile decomposition by-products. Furthermore, these co-ordination compounds are undergoing a defined decomposition mechanism and thus allowing a more accurate study of the respective metal oxide thin-film formation. Besides, the precursor solutions are processable without the requirement of additives or stabilizing agents, even under an inert atmosphere. Figure 19 illustrates the individual processing steps involved in the solution-processing of functional metal oxides by employing single-source molecular precursors.

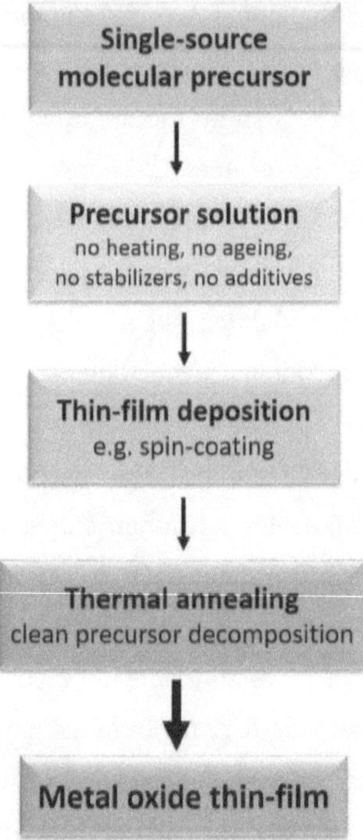

Figure 19 Schematic illustration of individual process steps involved in the solution-processing of metal oxide thin films employing single-source molecular precursors.

Furthermore, this approach enables the possibility of customizing the precursor ligand structure in order to obtain the desired chemical and physical properties, in order to achieve high quality metal oxide thin films. The desirable properties of the molecular precursors include air-stability, high solubility in aqueous and organic solvents, decreased PDA temperatures, a clean decomposition and the applicability of direct photo-patterning. Consequently, such well-defined single-source molecular precursors are highly promising materials for an efficient large-area solution-processing of functional metal oxides. The precursor compounds used within this book were purposely synthesized for the formation of dielectric and semiconducting metal oxide thin films and are depicted below. Details regarding the synthesis routes of these precursors are provided in section 6.

- Oximato complexes of Sn, In and Zn
- Nitro-functionalized malonato complexes of Al and Y
- Urea-nitrate complexes of Al, Y, Zn, In and Ga

Each of the utilized precursor types herein, possess particular advantages in comparison to the conventionally employed precursors for solution-based metal oxide thin films. The "oximato" complexes are composed of the methoxyiminopropionic ligands, coordinated to the respective metal cation. The schematic representation of the chemical structures of the utilized oximato complexes of tin, indium and zinc are shown in Figure 20.

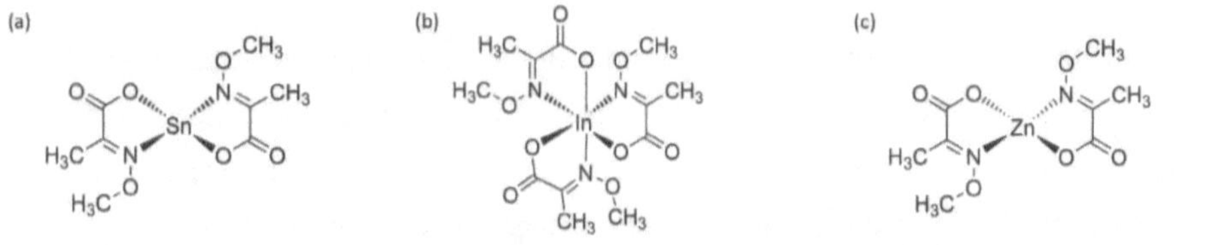

bis-(methoxyiminopropanoato)tin(II) (Sn-oximato) | tris-(methoxyiminopropanoato)indium(III) (In-oximato) | bis-(methoxyiminopropanoato)zinc(II) (Zn-oximato)

Figure 20 Schematic representation of the molecular structures of the oximato complexes of a) tin(II), b) indium and c) zinc.

The successful synthesis as well as various applications of the zinc and indium oximato precursors for the generation of well performing ZnO and IZO semiconductor thin films was previously reported by our group, demonstrating the potential of such defined precursor complexes for the formation of high quality metal oxide thin films towards the application in TFT devices.[73, 76, 77, 197-200] Although IZO and IGZO might be the preferred compositional choices for multinary metal oxide semiconductors, the expected scarcity of natural indium resources justify the need for alternative AOS such as ZTO.[201] However, the solution-processing of ZTO thin films enabled comparable TFT device performances to the indium based multinary AOS so far.[202-204]

During the course of this book the oximato precursor approach could be extended to the formation of high performing and indium-free, zinc-tin-oxide (ZTO), by the successful synthesis and structural characterization of bis[(methoxyimino)-propanoato]tin(II) (Sn-oximato). The synthesis of such a single-source molecular precursor for tin (II) is highly beneficial since the choice of the tin precursor has been mainly restricted to $SnCl_2$[205], which can create certain complications regarding the fabrication process, as discussed in section 4.1. Additionally, the zinc and tin (II) oximato precursor exhibit a very similar thermal decomposition behavior (TGA), which enables a homogenous start of the thermal decay in a narrow temperature window (125 - 150 °C), which results in a uniform distribution of the Zn, Sn(IV) and O species across the surface and in the depth of the films, confirmed by Auger electron spectroscopy.

As a result, an oximato precursor ratio of 7:3 of Sn:Zn represents the optimal composition and a processing temperature of 350 °C enables high performing ZTO-based TFTs exhibiting a μ_{sat} of 5.18 cm^2 V^{-1}s^{-1}, V$_{th}$ of 7.5 and I$_{on/off}$ of 6 x 10^8. The need for the higher Sn ratio can be attributed to the higher defect concentration of SnO_2 in comparison to ZnO, by using the oximato precursors, which was investigated by means of electron paramagnetic resonance (EPR) spectroscopy.

Further details regarding the precursor synthesis, the structural characterization and thermal decomposition behavior of the Sn-oximato precursor, as well as the material and electrical characterization of the generated SnO and ZTO thin films can be found in section 6.4.[206]

Recently, our group could successfully demonstrate the feasibility of the oximato precursors towards the novel approach of direct photo-patterning of high performance IZO and ZTO TFTs, employing the respective oximato precursors.[207] As a result, at a PDA temperature of 350 °C, the IZO and ZTO based TFT devices exhibit a μ_{sat} of 7.8 and 3.6 cm^2 V^{-1} s^{-1}, a threshold voltage V$_{Th}$ of 0.3 and 2.4 V as well as an on/off current ratio I$_{on/off}$ of 3.5 x 10^8 and 5.3 x 10^7, respectively.

4.2 Combustion synthesis of metal oxide dielectrics

Promising single-source molecular precursors typically start to decompose at moderate temperatures of about 200 °C, but unfortunately, organic residues from the ligand framework often remain in the metal oxide thin films at such low annealing temperatures, corrupting the overall electronic performance.[208] The complete decomposition of the ligand framework usually requires temperatures up to 450 °C.[209, 210] Thus a useful approach is the systematic introduction of reactive functional groups, like nitro or nitroso groups, into the ligand framework, which are enhancing the exothermic decomposition behavior of the precursor. As a result, lower processing temperatures can be realized for the complete decomposition of such combustible precursors. Figure 21 displays a schematic illustration of the reaction energy difference for the synthesis of metal oxides by employing conventional precursors as well as combustion precursors.[211, 212]

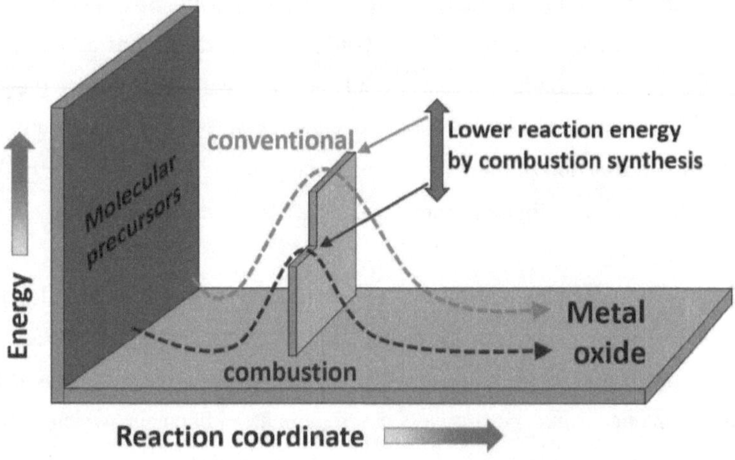

Figure 21 Schematic illustration of the required reaction energy for the conversion of the metal oxide precursor by using conventional precursors in comparison to combustion precursors.

In order to demonstrate the feasibility of this approach we have synthesized nitro-functionalized diethyl malonato complexes of aluminum and yttrium and investigated their potential towards low-temperature processing of dielectric Al_xO_y, Y_xO_y and compositional hybrid YAl_xO_y thin films. The chemical structures of the malonato complexes of aluminum and yttrium are presented in Figure 22.

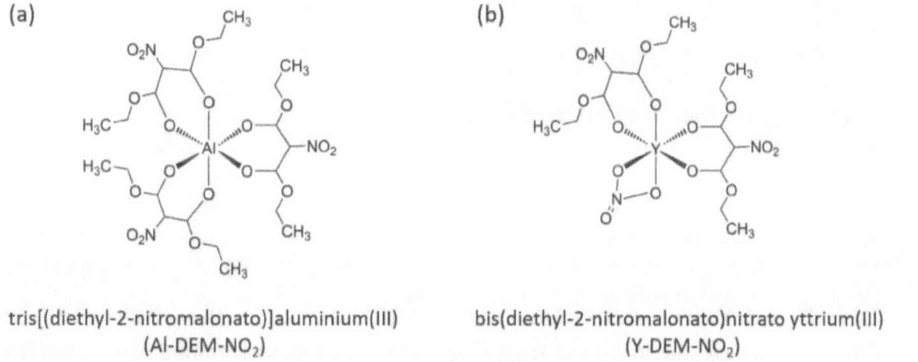

tris[(diethyl-2-nitromalonato)]aluminium(III)
(Al-DEM-NO₂)

bis(diethyl-2-nitromalonato)nitrato yttrium(III)
(Y-DEM-NO₂)

Figure 22 Schematic representation of the molecular structures of the synthesized malonato complexes of (a) aluminum and (b) yttrium, derived from the diethyl-2-nitromalonato ligand.

Tris[(diethyl-2-nitromalonato)]aluminum(III) (Al-DEM-NO₂) (Figure 22a) forms a homoleptic complex, with the aluminum central cation octahedrally coordinated by two oxygen atoms of the keto group from the dietyl-2-nitromalonate ligand. In contrast, bis[(diethyl-2-nitromalonato)]nitrato yttrium(III) (Y-DEM-NO$_2$) (Figure 22b) forms a heteroleptic complex, whereby one nitrate anion functions as ligand and coordinates bidentate to the yttrium cation. In accord with various chemical analysis the Y-DEM-NO$_2$ precursor might exhibit an octahedrally coordination of the yttrium, coordinated by two oxygen atoms of the keto group from each of the two diethyl-2-nitromalonate ligands as well as two oxygen atoms from the nitrate ligand.

The Al-DEM-NO$_2$ complex crystallizes as centimeter-long yellow needle-shaped crystals (Figure 23a) in the monoclinic space group C2/c. Despite numerous crystallization attempts the long needle like crystal structure obtained from the precursor is preventing a satisfactory crystal structure refinement, but nevertheless a preliminary refinement which substantiates the connectivity of the atoms is clearly demonstrating the octahedral coordination environment of the precursor compound (Figure 23b).

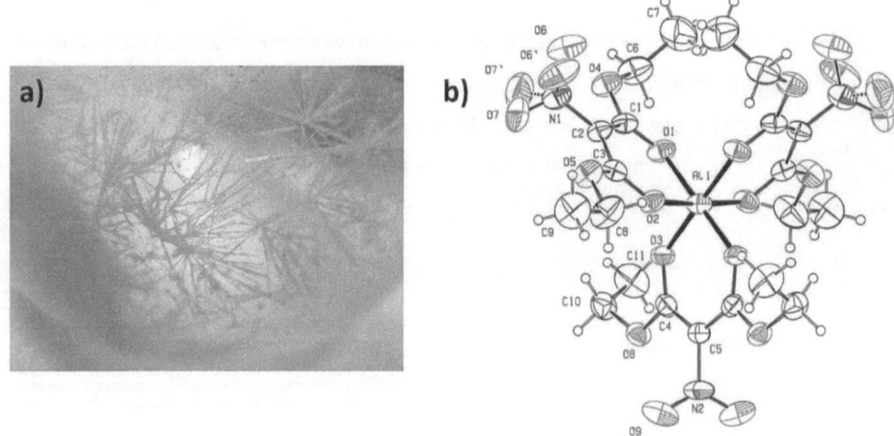

Figure 23 a) Photograph of needle-shaped crystals of Al-DEM-NO$_2$, b) ORTEP plot of the molecular structure of Al-DEM-NO$_2$. Vibrational ellipsoids are drawn with a probability of 30 %; O-Al-bond length (185-188 pm); O-Al-O-bond angle (90± 3°). Two NO$_2$ groups of the ligand show disorder in the oxygen positions.

Consequently, the application of tris[(diethyl-2-nitromalonato)]aluminum(III) (Al-DEM-NO$_2$) as combustible precursor for solution-processing of Al$_x$O$_y$ thin films was investigated and results in reduced processing temperatures in order to obtain good dielectric performances.

The integration of amorphous Al$_x$O$_y$ thin films, derived from the Al-DEM-NO$_2$ precursor, within capacitor devices exhibit dielectric behavior at a processing temperature as low as 200 °C, displaying a capacitance per unit area (C_i) of 41 nF cm^{-2}, a leakage current density (J) of 1.7 x 10^{-7} A cm^{-2} (at 1 MV cm^{-1}) and breakdown voltage (E_B) of 1.8 MV cm^{-1}. A slightly elevated PDA temperature of 350 °C enables the processing of capacitors exhibiting very good dielectric properties such as C_i = 86 nF cm^{-2}, J =9 x 10^{-10} A cm^{-2} (at 1 MV cm^{-1}) and E_B = 2.8 MV cm^{-1}.

The enhancement of the dielectric properties with increasing temperature can be attributed to the steady transformation of the intermediate aluminum oxo–hydroxy species into aluminum oxide, which was confirmed via X-ray photoelectron spectroscopy (XPS).

Furthermore, the fabrication of an IZO based solution-processed TFT, utilizing the Al_xO_y dielectric, processed at 350 °C, demonstrates good TFT performance characteristics with a μ_{sat} of 7.1 cm^2 V^{-1} s^{-1}, a V_{th} of 8.7 V and a $I_{on/off}$ of 1.4 x 10^5. Further details regarding the precursor synthesis, the structural characterization and thermal decomposition behavior of the Al-DEM-NO$_2$ precursor, as well as the material and electrical characterization of the generated Al_xO_y dielectric can be found in section 6.1.[213]

During the course of this book, we also investigated the potential of Y_2O_3 as gate dielectric by utilizing the previously introduced bis[(diethyl-2-nitromalonato)]nitrato yttrium(III) complex (Y-DEM-NO$_2$). The manufactured capacitor devices based on Y_xO_y depict dielectric behavior at PDA temperatures of 200-350 °C, with areal capacity values ranging from 21 up to 84 nF cm^{-2} at 10 kHz. The leakage current density for T = 300 and 350 °C amounts J < 1.0 x 10^{-9} A cm^{-2} at 1 MV cm^{-1} and the electrical breakdown field is E_{BD} > 2 MV cm^{-1}.

Additionally, a solution-processed TFT was fabricated, based on the Y_xO_y dielectric and derived from the Y-DEM-NO$_2$ precursor. The TFT device exhibits decent performance characteristics with a saturation mobility (μ_{sat}) of 2.1 cm^2 V^{-1} s^{-1}, a threshold voltage (V_{th}) of 6.9 V and an on/off current ratio ($I_{on/off}$) of 7.6 x 10^5. Further details regarding the precursor synthesis, the characterization and thermal decomposition behavior of Y-DEM-NO$_2$ as well as the material and electrical characterization of Y_xO_y can be found in section 6.2.[214]

Another work within this book was focusing on the optimization of the metal oxide gate dielectric by the generation of hybrid ternary oxides, due to the tunable electrical properties of such materials. Therefore, we decided to synthesis yttrium aluminium oxide (YAl_xO_y), due to its complementary properties, employing the previously established and advantageous single-source molecular precursors Al-DEM-NO$_2$ and Y-DEM-NO$_2$. The solution-processing of various Al:Y ratios, deposited via the spin-coating technique and annealed at 350 °C, resulted in YAl_xO_y dielectrics with excellent dielectric properties. All investigated compositions (10-50 mol-% Y) turned out to be amorphous throughout, even on the nanoscopic scale, representing a huge advantage in comparison to the pristine Y_xO_y dielectric, which starts to crystallize at 350 °C. Additionally, all YAl_xO_y samples exhibit a smoother surface roughness of $R_{RMS} < 0.2$ nm in comparison to the individual binary oxides. Interestingly, the inclusion of yttrium into the Al_xO_y host lattice reaches a saturation at an yttrium incorporation of 30 mol-%. Up to the saturation point, the compositions display lower amounts of carbonate (CO_3^{2-}) species (XPS) within the material and thus lower leakage currents of the final devices, in comparison to the Al_xO_y based capacitors. All YAl_xO_y-350 based capacitors possess a very low leakage current density of $J < 2$ x 10^{-9} A cm^{-2} at 1 MV cm^{-1}, showing no electrical breakdown up to electric fields of $E_B > 3.3$ MV cm^{-1}. In addition, all YAl_xO_y-350 compositions exhibit increased dielectric constants in comparison to the pure Al_xO_y-350 dielectric, whereby from an yttrium incorporation of ≥ 30 mol-%, the dielectrics display almost no frequency dispersion in the range of 10 Hz - 100 kHz. Regarding the investigated compositions, the 30-YAl_xO_y dielectric is exhibiting the best overall dielectric properties and consequently its feasibility as functional dielectric was demonstrated by the integration within TFT devices by using an active IZO semiconductor. The crucial TFT performance characteristics were evaluated, exhibiting reasonable values with a μ_{sat} of 2.6 cm^2 V^{-1} s^{-1}, V_{on} of -1.1 V, V_{th} of 12.4 V and $I_{on/off}$ of 1.8 x 10^7. Further details regarding the material and electrical characterization of the various YAl_xO_y compositions are provided in section 6.3.[215]

4.3 Aqueous combustion synthesis of metal oxide dielectrics

Another commonly practiced approach is the combustion synthesis of metal oxides based on a "fuel/oxidizer" reaction, which enhances the exothermic behavior, resulting in localized heating of the solution and consequently enabling a complete conversion of the precursor at reduced post-deposition annealing temperatures. Typically, metal nitrates are employed as "oxidizer" and urea or acetylacetone serve as "fuel" in such combustion synthesis processes.[216, 217] Thereby, the thermal decomposition initiates redox reactions between the nitrate anion and the urea or acetylacetone molecules, which enhances the conversion into the metal oxide while various nitrogen containing compounds are released as by-products.[218] Figure 24 illustrates the aqueous combustion synthesis approach, employing metal nitrates as "oxidizers" and urea as "fuel". In addition, the decomposition pathway of *in-situ* generated urea nitrate is demonstrated (Figure 24, right side).[219]

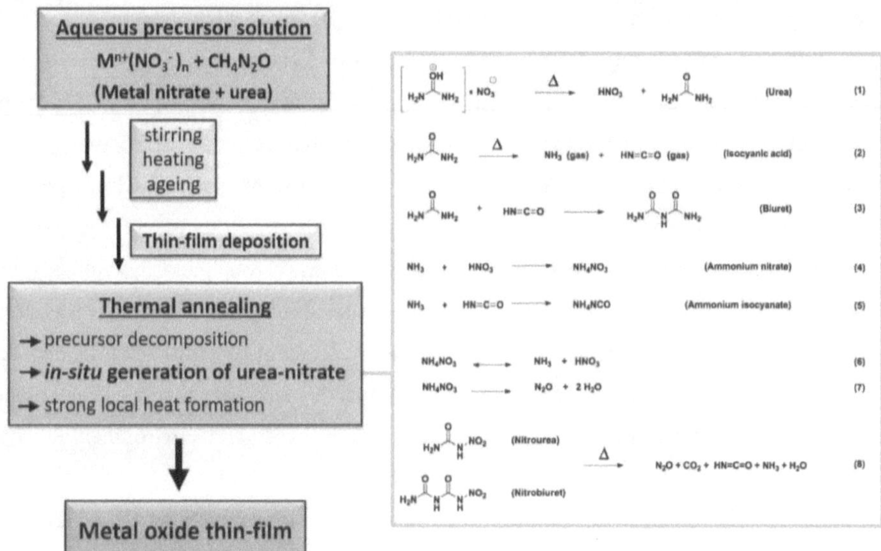

Figure 24 Schematic illustration of individual process steps involved in the aqueous solution-processing of metal oxide thin films employing metal nitrate precursors as oxidizers and urea as "fuel"; (right side) decomposition mechanism of urea-nitrate from 25 °C to 360 °C, determined by TG-FTIR-MS. Adapted with permission [219]. Copyright 2011, Elsevier B.V.

Consequently, research employing this strategy for the synthesis of functional metal oxides delivered very promising results.[220-225] From a technologically relevant point of view, the metal oxide thin-film formation from aqueous solutions represents a highly attractive solution-processing route for large-area applications. The motivation for using water as solvent are obviously relying on its advantageous properties such as representing an environmentally friendly and hydrocarbon-free solvent which is cost efficient and easy accessible. The aqueous combustion synthesis of aluminum oxide dielectrics was demonstrated for the first time by the Fortunato group.[18] As a result, DSC analysis revealed an initiating temperature below 200 °C for an optimized aqueous precursor solution and adjusted reaction parameters, suitable for the development of low-temperature solution-processed dielectrics. Furthermore, Al_xO_y thin films (10 nm) were prepared from an aqueous solution, deposited via the spin-coating technique and processed at 350 °C. Subsequently, the Al_xO_y dielectric was integrated in TFT devices, using a solution-processed GZTO semiconductor, exhibiting very good electrical performance characteristics. The TFTs exhibited a saturation mobility (μ_{sat}) of 1.3 cm^2 V^{-1} s^{-1}, on-voltage (V_{on}) of 0.5 V and a subthreshold swing (SS) of 0.3 V dec^{-1}. Thereby, the low operating voltages are attributed to the high areal capacitance of the employed high-k dielectric. Unfortunately, the leakage current and off current (I_{off}) were relatively high, which might be due to the low Al_xO_y film thickness of 10 nm. Within this study, also a comparison between water and 2-methoxyethanol was performed, investigating the influence of the employed solvent on the quality of the film formation. As a result, scanning electron microscopy (SEM) analysis revealed a difference in the Al_xO_y thin film morphology, whereby films generated from water-based precursors are smoother and more uniform in comparison to the use of the 2-ME based precursors. Complementary, AFM analysis display a surface roughness R_{RMS} of 0.3 nm and 0.9 nm for the water-based and 2-ME based precursor solution, respectively. Furthermore, the utilization of water-based precursor solutions result in comparatively thinner and more compact films, investigated by spectroscopic ellipsometry. Thus, this work surely proved the potential of aqueous combustion synthesis for low-temperature, low-cost and environmentally friendly metal oxide thin film fabrication.

As a fact, this approach already demonstrated its feasibility for solution-processed dielectrics as well as semiconductors so far, although a detailed decomposition mechanism of the combustion synthesis still needs to be clarified. Furthermore, aqueous solutions of metal nitrates, urea and other additives usually require ageing over a longer period of time, in order to evolve the full combustion effect and are susceptible to various processing issues, which renders the reproducibility of the combustion process complicated.[18]

Thus, we have chosen to combine the "oxidizer" (nitrate ions) and the "fuel" (urea ligands) directly in one discrete molecule and therefore purposely designed coordination compounds of aluminum, yttrium, indium, gallium and zinc. This should lead to the generation of binary and multinary metal oxide dielectric and semiconductor thin films, which would allow more control and reproducibility, enabling a systematic study of the respective metal oxide formation in an accurate manner. The exothermic decomposition of all investigated metal urea nitrate complexes could be confirmed by DSC, exhibiting sharp exothermic peaks in the temperature range of 100-225°C. In addition, all urea nitrate compounds undergo a clean decomposition during the combustion synthesis, releasing gaseous by-products, which were detected via TG-MS and could be assigned to ammonia (m/z + 17), water (m/z + 18), carbon monoxide (m/z + 28), nitric oxide (m/z + 30), isocyanic acid (m/z + 43), carbon dioxide (m/z + 44) and nitrogen dioxide (m/z + 46).The corresponding IR signals detected at the maximum of the Gram–Schmidt signal clearly confirm for all precursors the evolution of ammonia (NH_3: 965 cm^{-1}), isocyanic acid (HNCO: 2239 cm^{-1}), carbon dioxide (CO_2: 2310 and 2359 cm^{-1}) and nitric acid (HNO_3: 1629 and 1305 cm^{-1}), recorded at various decomposition stages between 200 °C and 350 °C. Furthermore, the employment of the respective urea-nitrate compounds allows an aqueous precursor route, by forming stable solutions without the requirement of any further stabilizing agent or additive. The proposed chemical structures of the urea-nitrate coordination compounds of aluminum and yttrium are shown in Figure 25.

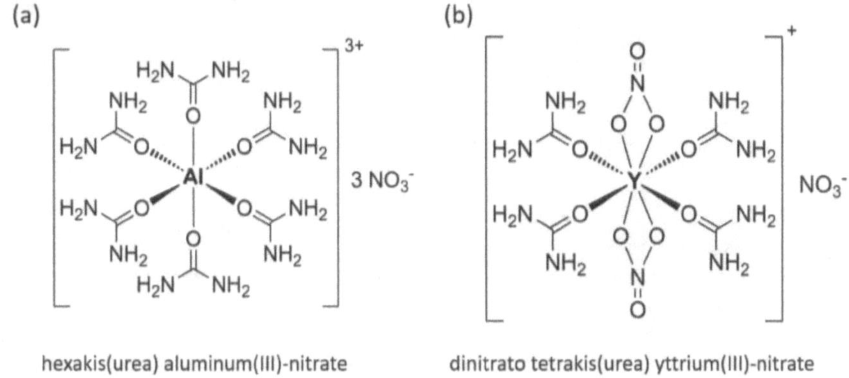

Figure 25 Schematic representation of the molecular structures of the urea nitrate coordination compounds of (a) aluminum and (b) yttrium.

The hexakis(urea)aluminum(III)-nitrate coordination compound $[Al(CH_4N_2O)_6][NO_3]_3$ (Figure 25a) forms a homoleptic complex, whereby the aluminum central cation is octahedrally coordinated by the single oxygen atom of the urea ligands, in accord with Pearson's hard/soft acid base concept. Thereby, Al-UN crystalizes in the space group $P2_1/n$, exhibiting white needle-shaped crystals. The long needle like crystal structure obtained from the precursor by crystallization enables a preliminary structural refinement. It clearly demonstrates the atom connectivity, displaying an octahedral coordination environment around the central yttrium atom. Comparable hexakis(urea) complexes of trivalent metal ions are also known e.g. for manganese[226] or chromium[227, 228].

Dinitrato tetra(urea) yttrium(III)-nitrate $[Y(CH_4N_2O)_4(NO_3)_2][NO_3]$ (Figure 25b) forms a heteroleptic complex, crystallizing in the space group $P\bar{1}$. Thereby, the nitrate anion functions as ligand and coordinates in a bidentate manner to the yttrium cation, while the neutrally charged urea molecules act as monodentate ligands and coordinate to the central metal atom by its oxygen atom. Here, the central metal is coordinated by four urea and two nitrate molecules, resulting in a double capped trigonal prism with coordination number eight for yttrium, shown in Figure 26.

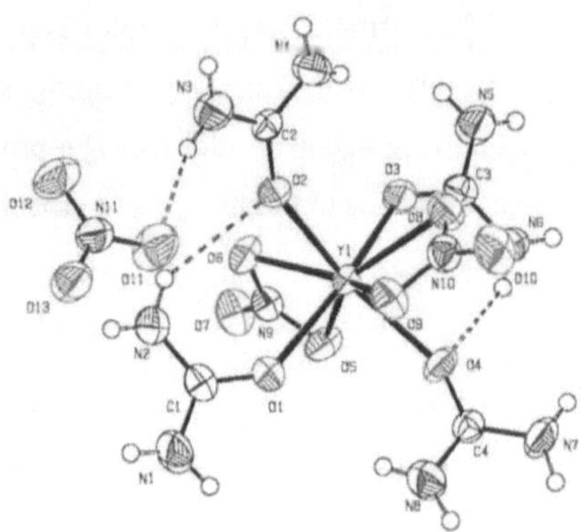

Figure 26 ORTEP plot of the molecular structure of Y-UN. Vibrational ellipsoids are drawn at the 50% probability. For the urea ligand the O–Y bond length are in the range 222–226 pm and for the nitrate ligand 243–249 pm; O–Y–O-bond angles are 90 ± 5° for the urea ligand.

The employment of the aluminum urea-nitrate precursor (hexakis(urea) aluminum(III) nitrate[218, 229] at an annealing temperature of 300 °C, results in amorphous Al_xO_y dielectrics with excellent dielectric properties, exhibiting a remarkable high areal capacitance of 216 nF cm^{-2} at 10 kHz, a leakage current density less than 1.0 x 10^{-9} A cm^{-2} at 1 MV cm^{-1} and an electrical breakdown field at E_{BD} = 3.0 MV cm^{-1}. Further details regarding the precursor synthesis, the characterization and thermal decomposition behavior of Al-UN as well as the material and electrical characterization of Al_xO_y can be found in section 6.2.[214]

Besides, it was possible to extent the combustion synthesis strategy to yttrium urea-nitrate for the formation of dielectric yttrium oxide (Y_xO_y). Thereby, the Y_xO_y thin films are showing decent dielectric properties at PDA temperatures T ≥ 300 °C, but suffer from a high degree of crystallinity, attributable to the strong exothermic decomposition of the precursor, leading to local hot-spots during the calcination, which enhance the crystallization. As a result, very thin films of the Y_xO_y dielectric, derived from the Y-UN precursor, show a high tendency of short circuits, rendering the whole process of oxide formation inefficient. Further information regarding the yttrium urea-nitrate coordination compound as well as the dielectric Y_xO_y thin films are also depicted in section 6.2.[214]

During the course of this thesis it was also possible to apply this approach to the generation of metal oxide semiconductor thin films, by the synthesis, characterization and employment of urea nitrate coordination compounds of indium, gallium and zinc. Figure 27 illustrates the molecular structures of the urea-nitrate precursors.

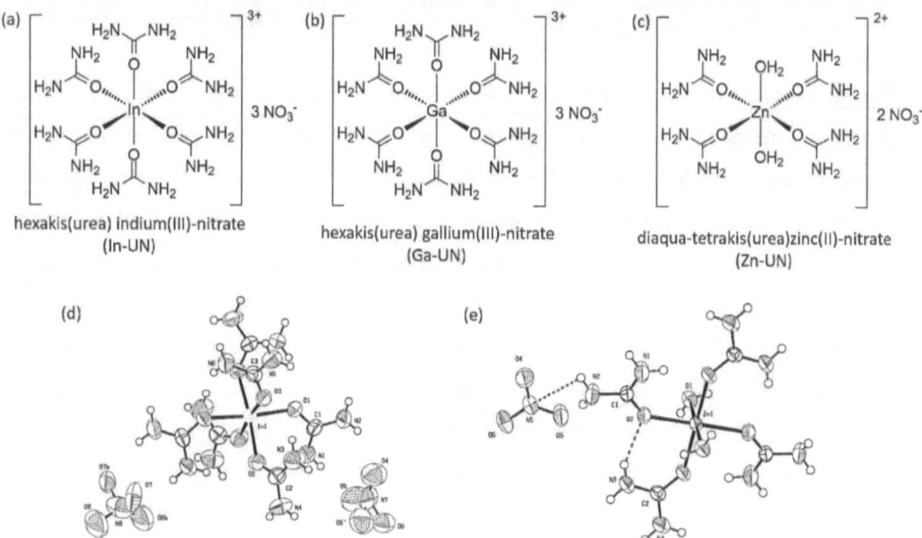

Figure 27 Schematic representation of the molecular structures of the urea nitrate coordination compounds of (a) indium, (b) gallium and (c) zinc and ORTEP plot of the urea-nitrate complexes of d) indium and e) zinc, obtained via single-crystal structure analysis and drawn at a 50 % probability. The structure of the indium complex displays two nitrate counter anions showing disorder in the oxygen positions.

The coordination compounds of indium and gallium are isotypic and crystallize in the space group C2/C, whereas crystals of the zinc complex exhibit the space group P21/n (Figure 27 d-e). Thereby, the indium and gallium precursor possess the general composition $[M(CH_4N_2O)_6][NO_3]_3$, as observed for the coordination compound of aluminum. In contrast, the zinc complex exhibits the general composition $[Zn(CH_4N_2O)_4(H_2O)_2][NO_3]_2$, as already reported.[230] Furthermore, complexes with the composition $[M(CH_4N_2O)_4(H_2O)_2][NO_3]_2$ were also found for copper, cobalt and nickel.[230, 231] It is possible to obtain water-free coordination compounds by simply using non aqueous reaction conditions as performed for copper and zinc complexes, resulting in tetrakis(urea) copper(II) nitrate $[Cu(CH_4N_2O)_4][NO_3]_2$ as well as hexakis(urea) zinc nitrate $[Zn(CH_4N_2O)_6][NO_3]_3$ complexes.[232, 233]

Details regarding the precursor synthesis, the structural characterization and thermal decomposition behavior of the precursor compounds used herein, as well as the material and electrical characterization of the generated IGZO semiconductor thin films are shown in section 6.5.[234]

The photo-processing technique for solution-processed metal oxides comprises of desired processing methods such as photo-annealing, photo-patterning or a combination of the two methods, whereby the photo-conversion to the desired final metal oxide can be achieved selectively as well as without the generation of significant heat. These methods enable the generation of high performance functional metal oxides, while simultaneously reducing the post deposition annealing (PDA) temperature to an amount which is compatible with the deposition on flexible substrates, such as plastic or paper, enabling the fabrication of flexible electronics. Thereby, it is highly advantageous to employ such techniques to reduce the processing temperatures below the glass transition temperature (T_g) of the flexible substrate, whereby the flexible substrates are thermally significantly less stable compared to e.g. glass or silicon substrates. Regarding the deposition on plastic substrates, potential candidates are polymers such as polyethelene terephthalate (PET), polyethylene naphthalate (PEN) (thermoplastics; semi-crystalline), polycarbonate (PC), polyethersulphone (PES) (thermoplastics; non-crystalline) as well as polyarylates (PAR), polycyclic olefin (PCO) and polyimide (PI) (high-T_g materials).[235] Among these polymers, polyimide exhibits a relatively high glass transition temperature of T_g ~350 °C and high resistance towards process chemicals, but most commercial PI exhibits low optical transparency, possessing a yellowish tint, which limits its application in transparent optical electronics. In contrast, PC, PES, PAR and PCO possess a relative high optical transparency, but they exhibit a poor resistance towards chemicals used during the conventional wet and dry etching processes used in photo-lithographic patterning. PET and PEN are transparent with a transmittance of ~85 % in the visible, exhibiting a reasonable resistance towards process chemicals and low water adsorption (~0.14 %), but allowing only processing temperatures of ~150 °C and ~200 °C, respectively.[235] Thus, the deposition of the functional metal oxide thin films on promising polymer substrates such as PEN and PET, for the fabrication of flexible electronics, allow processing temperatures of maximum 200 °C.

In recent years, photochemistry has emerged as a powerful method towards the formation of such solution-processed functional metal oxide thin films at low-temperatures.[68, 70, 151, 236-244] Photochemistry represents a field of chemistry, focusing on the effects caused by the interaction of light at different wavelengths (UV, visible and IR radiation) with chemical compounds. The first law of photochemistry known as the Grotthuss–Draper law states that photochemical processes only occur due to the absorption of electromagnetic radiation.[245] Due to the fact that light is a form of radiant energy, the absorption of light enables the electrons in a compound to promote from the ground state to an energetically excited state, initiating various chemical and physical processes. In other words, the excited-state energy is able to induce a variety of photochemical reactions leading to the dissociation of chemical bonds and simultaneously facilitating the formation of new bonds. If the employed irradiation wavelength (λ) corresponds to the respective bond dissociation energy, the chemical bond of the precursor cleaves and the transformation into the respective oxide can possibly occur without the need of thermal annealing. The required wavelength can be calculated according to the following equation:

$$E = \frac{h \cdot c}{\lambda}$$
Equation 5.1

E = energy [J], $\quad$ h = 6.626 x 10^{-34} J s^{-1}, $\quad$ c = 3 x 10^{8} m s^{-1}, $\quad$ λ = wavelength [m]

Thereby, radiation with appropriate wavelengths can be employed, depending on the absorption ability of the precursor system and its chemical components to the irradiation wavelength. In fact, the employment of various light sources, emitting radiation in the deep-UV range (λ < 260 nm), provide sufficient energy to initiate the photolytic dissociation of the chemical bonds typically present within the molecular structures of common organic precursor ligands as demonstrated in Table 2.

Table 2: Overview of chemical bonds typically present within as-deposited precursor solution thin films.[241]

Chemical bond	Dissociation energy [kJ mol^{-1}]	Dissociation energy [eV]	Wavelength λ [nm]
N≡N	941	9.75	127
C=O	805	8.34	149
N=N	631	6.54	190
C=C	607	6.29	197
O=O	498	5.16	240
O-H	464	4.81	258
H-H	436	4.52	274
C-H	413	4.28	290
N-H	393	4.07	304
C-O	358	3.71	334
C-C	347	3.60	345
C-N	305	3.16	392
O-O	204	2.11	586
N-O	201	2.08	595
N-N	160	1.65	748

Figure 28a summarizes the required steps for the UV-mediated low-temperature processing of such metal oxide thin films which includes the solution-processing of the precursor thin films, the photochemical activation of the functional metal oxide upon irradiation by UV light and finally the densification of the M-O-M network. Note, if the cleavage leads to the generation of smaller molecules or atoms, such as reactive oxygen species (e.g., radicals), they could further promote the elimination of organic residuals from the precursor as well as the condensation and densification of the amorphous metal-oxygen network (Figure 28b). Therefore, such a method represents a highly promising alternative processing technique to the thermal decomposition of precursor solutions into functional metal oxides. Moreover, photo-processing can be performed soley or in combination with thermal annealing depending on the degree of photolytic conversion of the precursor compounds upon irradiation.

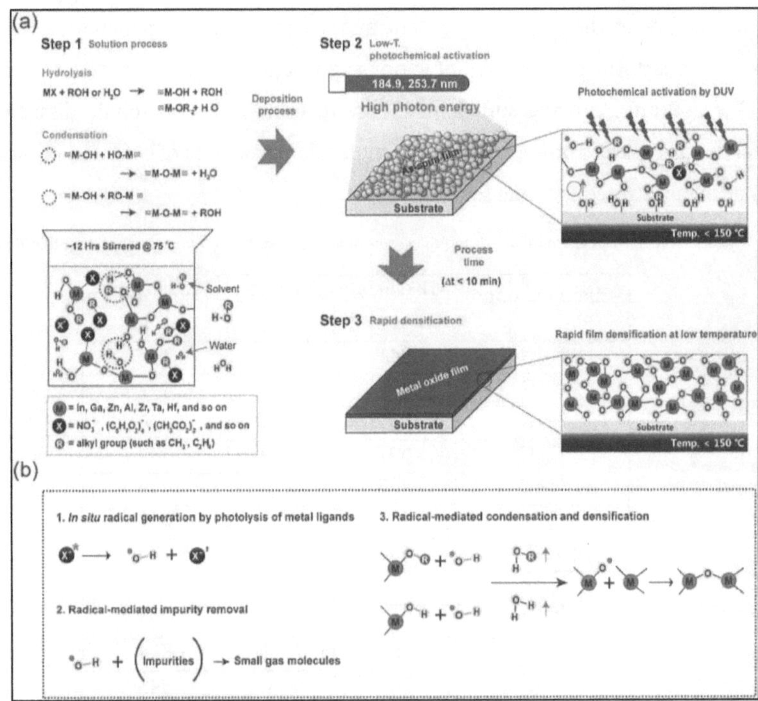

Figure 28 a) Schematic illustration of the general steps involved in the low-temperature photoactivation of various sol–gel metal oxide films; b) Mechanisms of the photoactivated solution process, proposed by Park et al.[240]. Copyright 2015, Wiley-VCH Verlag GmbH & Co. KGaA, Weinheim.

5.1 Advantages of the photo-induced decomposition over thermal decomposition in solution-processed metal oxide thin films

Recent investigations have demonstrated the benefits of employing UV irradiation for the solution-processing of functional metal oxide thin films in comparison with thermal annealing. Photo-processing results in an increased degree of M-O-M network formation, with enhanced density and an improved removal of impurities such as detrimental carbon species due to a highly selective bond cleavage. In-depth studies on aluminum oxide (Al_xO_y) thin films revealed the effect of UV irradiation on the solution-processing of such metal oxide thin films as shown in Figure 29, in the case of such a routinely used $Al(NO_3)_3$ precursor processed at (i) 150 °C without DUV irradiation, (ii) 150 °C with DUV irradiation and (iii) thermally annealed at 350 °C.[240] FTIR measurements, plotted as a function of the UV exposure time revealed the gradual removal of hydroxide species (3500 cm^{-1}) and ligand fragments (1400 and 1430 cm^{-1}) in the Al_xO_y thin films, by the employment of DUV irradiation (Figure 29a). It is remarkable that after only five minutes of UV exposure the FTIR spectrum is comparable to those of Al_xO_y thin films thermally annealed at 350°C, showing no remains of residual ligand species. Furthermore, time-of-flight secondary ion mass spectrometry (TOF-SIMS), confirmed a drastically reduced amount of carbon ions throughout the whole sample for the UV-treated thin films[240], which enhance the overall device performance, since trapped carbon species can cause an increased leakage current density or even short circuits when integrated into a device.

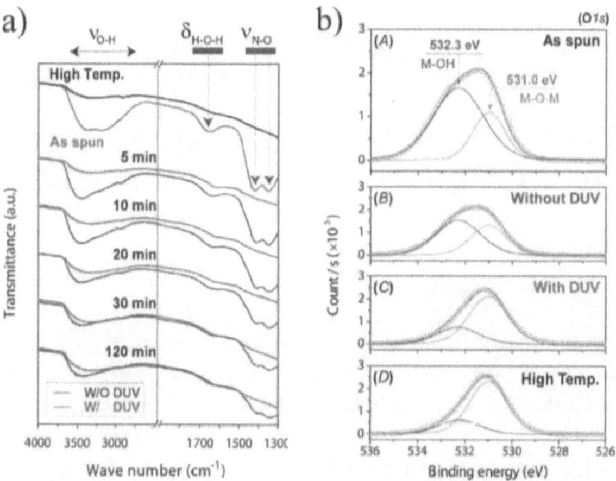

Figure 29 a) FTIR spectra of Al_xO_y thin films processed at 150 °C; without (blue line) and with (red line) DUV treatment, recorded at various UV exposure times and b) O 1s XPS core spectra of Al_xO_y thin films processed at different conditions: (A) treated at 150 °C for 5 min (B) without, (C) with DUV irradiation, and (D) thermally annealed at 350 °C for one hour. Adapted with permission[240]. Copyright 2015, Wiley-VCH Verlag GmbH & Co. KGaA, Weinheim.

X-ray photoelectron spectroscopy (XPS) reveal an increased transformation from the precursor state into the amorphous metal oxide network for thin films irradiated with UV (Figure 29b). The O 1s core spectra were deconvoluted into two peaks at 531.0 and 532.3 eV and are attributed to metal oxygen bonds (Al-O) and metal hydroxide species (Al-OH), which is in good agreement with other related studies.[246-248] The samples without UV treatment possess a significant amount of Al-OH species (532.3 eV), while the utilization of UV radiation results in comparatively higher amounts of Al-O (531.0 eV). In addition, thickness measurements display a decreased film thickness for the UV-treated Al_xO_y thin films, which can be attributed to a higher degree of densification of the amorphous metal oxygen (M-O-M) network as well as the enhanced removal of organic residues, induced by the UV irradiation.[240] Thereby, the low-temperature, UV-mediated solution process allows a more uniform and smooth film formation due to the slow drying of the solvent, resulting in decreased pore volumes within the final thin films. As a result, further densification can also be promoted by condensation reactions of the precursors due to a solvent-free film. Subsequently, dielectrics processed with DUV radiation possess enhanced dielectric performance characteristics, exhibiting lower leakage current densities and higher capacities per unit area, with a lower frequency dispersion in the range of 10^2 - 10^6 Hz, in comparison to samples thermally processed at 350 °C (Figure 30).

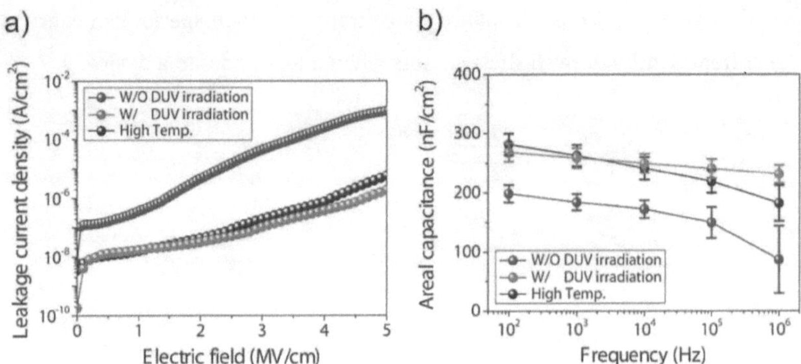

Figure 30 a) Leakage current density vs. e-field curves and b) areal capacitance vs. frequency curves of Al_xO_y dielectrics processed without DUV irradiation (blue), with DUV irradiation (red) and at a high annealing temperature (black). Adapted with permission[240]. Copyright 2015, Wiley-VCH Verlag GmbH & Co. KGaA, Weinheim.

Therefore, with all of the above mentioned benefits, the employment of UV radiation represents an attractive alternative processing technique towards low-temperature fabrication of thin-film devices. The reduction in PDA temperature enables the integration of these functional metal oxides into flexible electronic devices with a significantly reduced thermal budget. Thereby photochemistry could play a key role for future low-temperature solution-processing of metal oxides thin films.

5.2 Solution-based photo-processing of functional metal oxide thin films

Research in the last decade demonstrated highly promising results regarding the photo-processing of functional metal oxide thin films. In the year 2012, Kim *et al.* reported in a systematic study on various photo-processed semiconductor thin films such as IGZO and IZO, by utilizing metal salt precursors and employing a low-pressure mercury lamp in an inert N_2-atmosphere.[237] The employed mercury lamp emits radiation with a spectral distribution of λ = 253.7 nm/184.9 nm (90 %/10 %), accompanied by unintentional heating during the operation, which was measured to be T $\sim$ 150 °C. The effectiveness of the UV irradiation at such low temperatures was verified by reference samples processed soley at 150 °C, showing no or inferior electrical performance. The photo-processed TFT devices were fabricated by using a 35 nm thick Al_2O_3 gate dielectric, which was deposited via ALD. The Al_2O_3/IGZO and Al_2O_3/IZO based TFTs exhibit field-effect mobilities of 8.76 $\pm$ 0.98 and 4.43 $\pm$ 0.59 cm^2 V^{-1} s^{-1}, which are comparable or even higher than the devices thermally annealed at 350 °C. Finally, the process was transferred to commercially available flexible PAR substrates, by using the photo-processed IGZO semiconductor. TFT devices show excellent performance characteristics exhibiting a field effect mobility of 3.77 cm^2 V^{-1} s^{-1}, threshold voltage of 2.70 $\pm$ 0.47 V, subthreshold swing of 95.8 $\pm$ 20.8 mV dec^{-1} and current on/off ratio of 10^8.

In another similar work, In_2O_3 and IZO semiconductor thin films (15 nm) were generated by using the respective nitrate precursors and employing an UV-source with a broad emission spectra (200−300 nm), with the main emission peak at 250 nm. This study demonstrated the enhancement of the saturation mobility with increasing UV irradiation time, reaching a maximum after 15 minutes, displaying outstanding saturation mobilities of 34.44 and 33.14 cm^2 V^{-1} s^{-1} for In_2O_3 and IZO, respectively.[243]

In the year 2016, the Fortunato group transferred the photo-processing of functional metal oxide thin films to the generation of aluminum oxide dielectrics. The authors reported on the solution-based combustion synthesis of Al_xO_y thin films, employing an aluminum nitrate precursor, processed at 180 °C for 30 minutes and assisted by far-ultraviolet (FUV) radiation with a maximum emission at λ = 160 nm. TFTs based on the photo-processed Al_xO_y dielectric as well as a sputter-deposited IGZO semiconductor displayed excellent performance characteristics such as a low operation voltage, low turn-on voltage of V_{on} = - 0.12 ± 0.05 V and low subthreshold slope of SS = 0.11 ± 0.01 V dec^{-1}, exhibiting a very good reproducibility and stability over time.[242]

Recently, Subramanian *et al.* reported on the photo-processing of printed aluminum oxide thin film dielectrics.[68] Herein, precursor inks were prepared from aluminum nitrate and processed using UV radiation with a main emission peak at λ > 350 nm for 20 minutes and subsequent post-deposition annealing at 250 °C for two hours. As a result, a leakage current density of J = 5 x 10^{-6} A cm^2 and breakdown voltage of 5.1 MV cm^{-1} was achieved, representing an improvement in comparison to samples processed only at 250 °C and without UV exposure. Furthermore, Al$_x$O$_y$-based TFTs combined with an In$_x$O$_y$ semiconductor exhibited an average linear mobility of 12 ± 1.6 cm^2 V^{-1} s^{-1}, a minimal hysteresis and low subthreshold swing. Moreover, the TFT operational stability was improved by the use of UV irradiation.

The studies carried out during this book conducted several investigations on solution-processed, low-temperature, DUV-mediated amorphous metal oxide dielectrics. The utilized light-source emits DUV radiation with a peak intensity at λ = 160 nm and was operated under an inert argon atmosphere for the processing of the respective films. In addition, an integrated air cooling system ensured temperatures of T < 50 °C, allowing to investigate the pristine effect of the DUV treatment. Thereby, the previously introduced single-source molecular precursors tris(diethyl-2-nitromalonate)aluminum(III) (Al-DEM-NO$_2$ and bis[(diethyl-2-nitromalonato)] nitrato yttrium(III) (Y-DEM-NO$_2$) were investigated towards their potential be photo-processed, due to their strong absorption at 250 nm (Figure 31). Besides, the UV-Vis spectra of both precursors display a local absorbance maximum at ~ 310 nm, which is in accord with the yellow color of the precursor compounds.

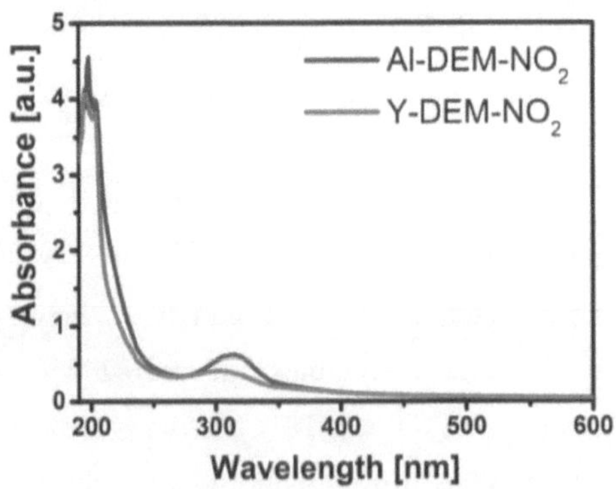

Figure 31 UV-Vis spectra of the single-source molecular precursors Al-DEM-NO2 and Y-DEM-NO2.

Surprisingly, the Y-DEM-NO$_2$ precursor does not undergo photolytic decomposition in an efficient manner, by employing UV radiation with a main emission peak at λ = 160 nm, resulting exclusively in electrical short circuits when depositing a single layer of the yttrium precursor solution. The employment of three layers results in very thick (329 nm) Y$_x$O$_y$ dielectrics, exhibiting degraded electrical properties in comparison to thermally annealed samples (T $\geq$ 250 °C), possessing a dielectric constant of k = 2.9 and leakage current density of 5.8 x 10^{-8} A cm^{-2} (at 1 MV cm^{-1}). The very high film thickness and degraded electrical characteristics are attributable to an incomplete metal oxide formation as well as enhanced void formation during the decay of the precursor compound. Interestingly, further investigations revealed that irradiation with main emission peaks at 254 and 185 nm leads to an improved bond cleavage of the precursor which is subject to future investigations. Hence, the ability of unique ligand fragments for a given metal oxide precursor to undergo complete decomposition upon irradiation with specific wavelengths and the quality of the resultant metal oxide makes it an interesting subject for future research.

In contrast, the use of the Al-DEM-NO$_2$ precursor enables the fabrication of DUV-mediated high performance dielectric Al$_x$O$_y$ based capacitor devices at processing temperatures as low as T = 150 °C (solvent evaporation prior to UV exposure). The photo-processing results in capacitor devices with excellent dielectric properties, exhibiting a dielectric constant of k = 9.0, which corresponds to a capacitance per unit area of 153 nF cm^{-2}. The leakage current density and breakdown voltage amount 1.7 x 10^{-9} A cm^{-2} (at 1 MV cm^{-1}) and 4.1 MV cm^{-1}, respectively, which is especially remarkable considering a single layer deposition of the precursor solution, possessing a final oxide thickness of 52 nm. The XPS O1s core spectra reveal that the DUV-mediated, low-temperature solution-processed Al$_x$O$_y$ thin films contain no remains of any organic constituents, arising from an incomplete precursor decomposition, comparable to thin films thermally annealed at T $\geq$ 300 °C. Additionally, an remarkably high degree of film smoothness of R$_{RMS}$ = 0.14 nm was observed for the DUV-processed Al$_x$O$_y$ thin films, which is much smoother than for samples thermally annealed at 350 °C. This is highly desirable since it enables a smooth dielectric/semiconductor interface formation in a TFT device. Subsequently, low leakage current densities of the final devices could be realized, due to small interface trap densities. Hence, the employment of the Al-DEM-NO$_2$ precursor irradiated at λ = 160 nm, enables the fabrication of low-temperature, solution-processed high performing Al$_x$O$_y$ thin films, promoting their application in flexible electronics. Complete details of the processing parameters as well as the material and electronic characterization can be found in section 6.3.[215]

The cumulative part of this book is divided into two main sections. The first section consists of the publications associated with the synthesis and full structural elucidation of single-source molecular precursors of aluminum and yttrium for low-temperature, solution-processing of the respective binary and ternary metal-oxide dielectrics (Al_xO_y, Y_xO_y and YAl_xO_y) as well as their integration in capacitor and TFT devices. Thereby, the novel and advantages malonato complexes of aluminum and yttrium, exhibiting the nitro functionalized diethyl molanate ligand as well as the respective urea-nitrate coordination compounds were employed, enabling the combustion synthesis of aluminum and yttrium oxide thin films.

The second section consists of publications associated with the synthesis and characterization of single-source molecular precursors for the formation of multinary n-type semiconductors, namely ZTO and IGZO, employing oximato complexes of Sn and Zn and urea-nitrate coordination compounds of In, Ga and Zn, respectively.

6.1 Synthesis, dielectric properties and application in a thin film transistor device of amorphous aluminum oxide Al_xO_y using a molecular based precursor route

Aluminum oxide (Al_2O_3) is a promising candidate for the future use as gate dielectric in TFT devices, due to its relative high dielectric constant ($k \approx 10$), wide band gap ($E_g \approx 8.8$ eV) and the ability to retain the amorphous phase at post-deposition annealing (PDA) temperatures T < 500 °C.

In this publication we demonstrate the formation of amorphous aluminum oxide thin films by a single-source molecular precursor approach, utilizing the coordination compound tris[(diethyl-2-nitromalonato)]aluminum(III) (Al-DEM-NO$_2$). The Al-DEM-NO$_2$ precursor compound was systematically characterized by spectroscopic techniques (NMR/IR) as well as by single-crystal structure analysis. Furthermore, the precursor was investigated towards its thermal decomposition behavior, by means of differential scanning calorimetry (DSC) and thermogravimetry coupled with mass spectrometry and infrared spectroscopy (TG-MS/IR). The precursor decomposes under oxygen and without the need of additional additives and stabilizers, by a thermal combustion into amorphous aluminum oxide thin films with a very low surface roughness of R_{RMS} ~0.3 nm. The amorphous phase was confirmed by means of transmission electron microscopy (TEM), showing no crystalline domains even on the nanoscopic scale throughout the entire sample. Furthermore, the material retains the amorphous phase over a broad temperature range, starting to crystallize at T ≥ 800 °C and forming α-Al_2O_3. Thus, solution-processing of the Al-DEM-NO$_2$ precursor results in the formation of smooth, dense and crack-free films, which are converted into amorphous Al_xO_y thin films after further calcination. The integration of amorphous Al_xO_y thin films within capacitor devices exhibit dielectric behavior at post-deposition annealing (PDA) temperatures between 200 and 350 °C. As a result, higher processing temperatures result in enhanced electrical performances, whereby capacitors processed at 350 °C display a capacity per unit area of 86 nF cm^{-2} , leakage current density of 8.9 x 10^{-10} A cm^{-2} (at 1 MV cm^{-1}) and electrical breakdown field of 2.8 MV cm^{-1}. The improvement of the electrical performance at higher temperatures can be attributed to the stepwise conversion of the intermediate aluminum oxo–hydroxy species into aluminum oxide, which was confirmed by X-ray photoelectron spectroscopy (XPS).

Finally, the feasibility of the solution-processed Al_xO_y thin films as gate dielectric in TFT devices was demonstrated. The fabrication of the TFT was carried out by employing the Al_xO_y dielectric and a solution-processed indium zinc oxide (IZO) semiconductor, which was subsequently annealed at 350°C. The manufactured Al_xO_y based TFTs demonstrate good TFT performance characteristics with a μ_{sat} of 7.1 cm^2 V^{-1} s^{-1}, a V_{th} of 8.7 V and an $I_{on/off}$ of 1.4 x 10^5.

Journal of
Materials Chemistry C

PAPER

Cite this: *J. Mater. Chem. C*, 2019,
7, 1048

Synthesis, dielectric properties and application in a thin film transistor device of amorphous aluminum oxide Al_xO_y using a molecular based precursor route†

Nico Koslowski,[a] Shawn Sanctis, [a] Rudolf C. Hoffmann,[a] Michael Bruns‡[b] and Jörg J. Schneider *[a]

Amorphous aluminium oxide thin films are accessible by a molecular single source precursor approach employing the coordination compound tris[(diethyl-2-nitromalonato)]aluminium(III) (Al-DEM-NO₂). The precursor decomposes by thermal combustion under oxygen without the need of an additional additive into amorphous aluminium oxide films at 350 °C with a very low surface roughness of about 0.3 nm. Solution processing of the precursor results in the formation of smooth, dense and crack-free films, which are converted into amorphous Al_xO_y thin films after further calcination. Amorphous Al_xO_y thin films integrated within a capacitor device exhibit dielectric behavior in the temperature range between 200 and 350 °C, with areal capacity values between 41 and 86 nF cm⁻² and leakage current densities ranging from 1.7×10^{-7} to 8.9×10^{-10} A cm⁻² (at 1 MV cm⁻¹) whereas breakdown voltages increase from 1.82 to 2.79 MV cm⁻¹ in the temperature regime from 200 to 350 °C. The increase in performance at higher temperatures can be attributed to the stepwise conversion of the intermediate aluminium oxo–hydroxy species into aluminium oxide which is confirmed by X-ray photoelectron spectroscopy (XPS).

Received 14th September 2018,
Accepted 18th December 2018

DOI: 10.1039/c8tc04660c

rsc.li/materials-c

Introduction

High-*k* dielectrics have received much attention as an important component in thin film transistors (TFTs) due to their high dielectric constant and the possibility to obtain a smooth dielectric/semiconductor interface.[1–5] Typically TFTs are manufactured using vacuum based processing techniques like atomic layer deposition (ALD), pulsed laser deposition (PLD), chemical vapor deposition (CVD) and radio frequency magnetron sputtering.[6–10] The use of amorphous metal oxides (AMOs) as gate dielectrics has enabled alternative processing techniques like deposition from solution. This allows coating of large-area substrates as well as the use of printing techniques.[2,11–21] Amorphous metal oxides like ZrO_2, HfO_2, Ta_2O_5, Y_2O_3 or TiO_2 (high-*k* dielectrics) are potential candidates to replace SiO_2 as gate dielectric.[22–29] Another promising high-*k* dielectric material for solution based processing techniques is aluminium oxide.[2,5,13–21,30–36] In a systematic investigation In_2O_3 and InZnO as semiconductors for TFTs as well as the dielectrics Al_2O_3, ZrO_2, Y_2O_3 and TiO_2 have been deposited by spin-coating from solution followed by subsequent calcination at 300 °C. Using the dielectric Al_2O_3 the best overall performance could be achieved. According to these studies there is a minor accumulation of charge carriers within the bulk dielectric as well as a small trap density at the semiconductor/dielectric interface.[13] Keszler *et al.* demonstrated that films of Al_2O_3, which are deposited *via* a sol–gel-process and calcined at 500 °C, exhibit comparable dielectric properties to those obtained from atomic layer deposition.[37]

Lately there have already been some attempts on low temperature solution processing of aluminium oxide aiming towards the application in flexible electronics. Carlos *et al.* prepared aluminum oxide films by solution combustion synthesis (SCS) at 180 °C under far-ultraviolet (FUV) irradiation for 30 minutes. The FUV irradiation improves film densification, resulting in a lower frequency dispersion and leakage current density. The application of the Al_xO_y dielectric in TFTs using IGZO as the semiconductor exhibited a great electrical performance and good reproducibility, being compatible for the use of flexible substrates.[33]

In the majority of studies, conventional aluminium salts like aluminium nitrate[13,15–17,19,30,31,33–36] or aluminium chloride[2,5,14,32,38]

[a] *Fachbereich Chemie, Eduard-Zintl-Institut für Anorganische und Physikalische Chemie, Technische Universität Darmstadt, Alarich-Weiss-Str. 12, 64287 Darmstadt, Germany. E-mail: joerg.schneider@ac.chemie.tu-darmstadt.de*
[b] *Institute for Applied Materials (IAM-ESS) and Karlsruhe Nano Micro Facility (KNMF), Hermann-von-Helmholtz-Platz 1, Karlsruhe Institute of Technology (KIT), 76344 Eggenstein-Leopoldshafen, Germany*
† Electronic supplementary information (ESI) available. See DOI: 10.1039/c8tc04660c
‡ XPS studies.

are used as precursors for the formation of Al_xO_y dielectrics. Although capacitors based on these precursors have shown good applicability, they seem to possess some drawbacks like the need for addition of further promoters or stabilizers. The use of such additives can have an uncertain impact on the film formation and the resulting electrical properties.[39] Regarding aluminium nitrate based precursor solutions, mono ethanolamine (MEA) is often added as a stabilizing agent.[13,15,40] Mono ethanolamine and other stabilizers can also function as ligands and are able to form complexes with the metal oxides in which components of those are found to be present on the resulting metal oxide film.[39] As a consequence this could lead to higher decomposition temperatures, in order to remove these residual organic complexes.

In the case of metal chlorides, $e.g.$ the formation of acidic by-products like HCl vapours during the calcination of the precursor thin films can cause an increased degree of porosity, resulting in deteriorated interface properties. Also, residual Cl^- ion traces are found to be present within the final ceramic and might act as charge traps at the interface.[41–43] All these undesired effects can have a drastic impact on the overall performance of the final dielectric material.

Herein, we report on the synthesis and structural elucidation of the new single source molecular precursor tris[(diethyl-2-nitromalonato)]aluminium(III) (Al-DEM-NO$_2$) for the preparation of dielectric aluminium oxide by solution processing. The precursor is highly soluble (>20 wt%) in 2-methoxyethanol, which is a conventional solvent for spin-coating. It decomposes under thermal combustion without the additional need of any additive at a fairly low temperature to yield smooth, dense and crack free films displaying a high capacitance of amorphous aluminium oxide. The precursor is thus suitable for the generation of such dielectric thin films, which can serve as a gate dielectric in TFTs which is shown in here, too.

Experimental section

Synthesis and characterization of tris[(diethyl-2-nitromalonato)]aluminium(III), Al-DEM-NO$_2$ 1

4.31 g (21 mmol) of 2-nitro-diethylmalonate was dissolved in 40 mL methyl $tert$-butyl ether (MTBE), whereby the reaction vessel was placed in a water bath at 5 °C. 0.510 g (7 mmol) of trimethylaluminium was added under stirring, whereby a steady gas evolution as well as a yellow colouring of the solution was observable. The solution was stirred for 18 h. Subsequently the yellow solution was concentrated by rotary evaporation leaving a yellow oil. The oil was dissolved in 25 mL dichloromethane (DCM) and filtered through a 0.2 μm polytetrafluoroethylene (PTFE) syringe filter. Finally, 200 mL of n-pentane was added to yield a clear yellow solution. After 48 h at 5 °C, needle shaped yellow crystals were formed. The obtained yellow crystals were finally dried in vacuum. The yield was 3.81 g (84.89%). Elemental analysis: C 39.45%, H 4.92% and N 6.51%. Calculated for $AlC_{21}H_{30}O_{18}N_3$: C 39.44%, H 4.73% and N 6.57%. ^{1}H-NMR (500 MHz, [d$_1$]CDCl$_3$): δ = 1.35 (t, –CH$_3$); 4.33 (m, –CH$_2$) ppm. ^{13}C(^{1}H)-NMR (500 MHz, [d$_1$]CDCl$_3$): δ = 14.10 (–CH$_3$);

63.97 (–CH$_2$); 109.99 (–C–NO$_2$); 168.63 (–C=O) ppm. ^{27}Al-NMR (500 MHz, [d$_1$]CDCl$_3$): δ = 0.97 (Al–O) ppm. Mass spectrum m/z = 639.5 (1%) (Al[C$_7$H$_{10}$O$_6$N]$_3$), 435.4 (100%) (Al[C$_7$H$_{10}$O$_6$N]$_2$) (see ESI† for mass spectrum).

Aluminium oxide formation from 1 and capacitor device fabrication

A precursor solution was prepared by dissolving 20 wt% of Al-DEM-NO$_2$ 1 in 2-methoxyethanol followed by subsequent sonication (240 W/L) for 20 minutes and filtering of the solution through a 0.2 μm polytetrafluoroethylene (PTFE) syringe filter. Capacitors have been fabricated using ITO-coated glass substrates (140 nm, OLED-grade). At first, the substrates were cleaned in deionized water, acetone and isopropanol, for 10 minutes each, by using an ultrasonic bath. Thereafter, a 100 nm gold layer was deposited with help of a shadow mask, using the sputter technique, in order to enable the electrical contact with the ITO layer. Then the substrates were exposed to air-plasma for 2 minutes to enhance the hydrophilicity. Subsequently the clear precursor solution was spin-coated (20 s; 3000 rpm) on the substrate and annealed at different temperatures (200, 250, 300 and 350 °C). This procedure was repeated three times. The thickness of the layer of 122–170 nm was determined by spectroscopic ellipsometry. As the top electrodes a 50 nm gold layer as well as a 10 nm interlayer titanium were sputtered on the Al_xO_y film with the help of another shadow mask. Al_xO_y thin films were annealed at 200, 250, 300 and 350 °C and are abbreviated as Al_xO_y-200, Al_xO_y-250, Al_xO_y-300 and Al_xO_y-350, respectively.

Thin film transistor fabrication

An indium zinc oxide (IZO) semiconductor was prepared by employing the previously established oximato precursors compounds[55–57] to fabricate the IZO thin film semiconductor. Solutions (1 wt%) of the indium and zinc precursors were prepared in 2-methoxyethanol (In : Zn, 6 : 4), spin coated at 2500 rpm for 20 seconds onto the Al_xO_y dielectric layer and finally annealed at 350 °C. Three iterations of the coating procedure were performed to obtain a desired film layer thickness of ~12–15 nm. For the final device fabrication metallic gold source–drain (W/L = 2 mm/150 μm) electrodes were sputtered through a metal based shadow mask (2 rectangular areas of 2 mm × 1 mm which are separated by a distance of 150 μm) onto the IZO semiconductor.

Materials characterization

Transmission electron microscopy (TEM) was performed using a Tecnai F20 (FEI) system, with an operating voltage of 200 kV. The samples have been prepared on lacey carbon coated copper grids. X-ray photoelectron spectroscopy (XPS) measurements were performed on a K-Alpha XPS system (Thermo Fisher Scientific, East Grinstead, UK). The monochromated Al Kα X-ray source was used at a spot size of 400 μm. All spectra are referenced to the C1s peak of hydrocarbons at 285.0 eV. Thermogravimetric analysis (TGA) coupled with infrared (IR) spectroscopy and mass spectrometry (MS) was carried out using TG 209N1 (Netzsch) coupled with a Nicolet iS10 spectrometer (Thermo Fisher Scientific) and Aeolos QMS 403C (Netzsch). The samples have been measured in an oxygen as well as an argon-atmosphere with a heating rate of 5 K min^{-1}, in the range of 50–600 °C, in an aluminium crucible. Differential scanning calorimetry (DSC) was carried out with STA 449 F3 Jupiter (Netzsch). X-ray diffraction (XRD) measurements were carried out using a MiniFlex 600 (Rigaku), using Cu-Kα-radiation with 600 W in the Bragg–Brentano geometry. Atomic force microscopy (AFM) was performed using a CP-II (Bruker-Veeco) with a silicon-Cantilever (T 3.75 µm, L 125 µm, W 35 µm, f_0 300 kHz, k 40 N m^{-1}) in the tapping mode at 1 Hz. Spectroscopic ellipsometry was performed using a Woolham M-2000 V spectrometer (spectral range 370–1690 cm^{-1}) using the completeEASE software (version 6.29). X-ray reflectometry (XRR) measurements have been performed using a Rigaku SmartLab with Cu source, rotating anode (K-Alpha 1 and 2) and parallel beam soller slits at 5°. The width of the incident slit and receiving slits are 0.1 mm and 0.2 mm, respectively. IR-spectroscopy was carried out on a Nicolet 6700 (Thermo Fisher Scientific). The samples have been measured in attenuated total reflection (ATR) without further preparation. NMR-spectroscopy was performed at 500 MHz using a DRX500 spectrometer (Bruker BioSpin GmbH). Mass-spectrometry was performed with MAT95 using a B-E sector mass spectrometer.

Electrical characterization

Impedance measurements have been carried out using a ModuLab MTS System (Solartron Analytical Ltd) in a glovebox under inert conditions. The instrument was equipped with a probe station (Cascade Microtech, Inc). Impedance Measurements have been carried out in the frequency range of 10 Hz to 100 kHz with amplitudes of 500 mA. The measurements of the breakdown voltage were performed on a B1500A semiconductor device analyzer (Agilent). TFT transfer characteristics were determined at V_{ds} = 15 V with an HP 4155A semiconductor parameter analyzer (Agilent) in a glove box under exclusion of air and moisture in the dark. Charge-carrier mobility in the saturation region (μ_{sat}) was calculated via the transconductance method and the threshold voltage (V_{th}) was derived from a linear fitting of the square root of the source–drain current ($\sqrt{I_{ds}}$) as a function of the gate–source voltage V_{gs}.

Results and discussion

Precursor synthesis and Al$_x$O$_y$ fabrication and characterization

Molecular single-source precursors can be specifically tailored, whereby the structure of the ligand framework around the metal cation tunes the chemical and physical properties of the precursor, resulting in comprehensible decomposition pathways. This enables a unique route for obtaining functional metal oxides. In this work we extended our single source precursor approach to the synthesis of a selectively designed aluminium coordination compound with a nitro-functionalized diethyl malonate ligand, for the formation of dielectric aluminium oxide (Scheme 1). Therefore, we have studied systematically

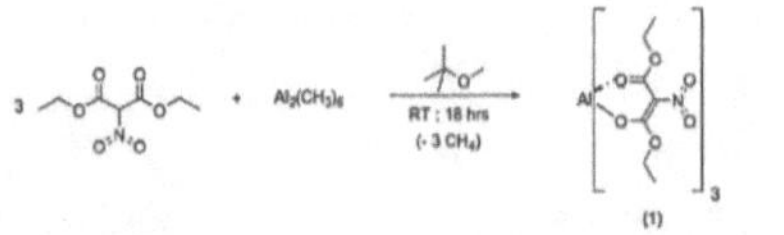

Scheme 1 Reaction scheme for the synthesis of tris[(diethyl-2-nitro-malonato)]aluminium, Al-DEM-NO$_2$.

and extensively possible synthesis routes for tris[[(diethyl-2-nitromalonato)]aluminium **1**. Typically, various aluminium alkoxides are used as aluminium source for such reactions which are able to form oligomers, so the complete substitution of all aluminium alkoxide groups takes place under carefully chosen reaction conditions. Additionally, the purification of the product seems to be difficult using alkoxides. In this work we used trimethylaluminium as the aluminium source, showing a good handling under cooling and inert conditions, leading to a pure product with satisfying yield. The nitro-functionalized diethyl malonate ligand exhibits a relatively high acidity and should react faster and completely.

Tris[[(diethyl-2-nitromalonato)]aluminium crystallizes as centimeter-long yellow needle-shaped crystals in the monoclinic space group $C2/c$ and is an isotype of tris[[(dimethyl-malonato)]-aluminium **1**, in which the malonate ester backbone is not functionalized.[44] In both metal complexes aluminium is coordinated octahedrally by two oxygen atoms of the keto group. Despite numerous crystallization attempts the long needle like crystal structure obtained from the precursor had so far prevented a satisfactory crystal structure refinement; however, a preliminary refinement clearly demonstrates the connectivity and the octahedral coordination environment of the precursor compound (Fig. S1, ESI†). The octahedral structure of the precursor compound is additionally confirmed by ^{27}Al-NMR spectroscopy, displaying a single sharp peak at 0.97 ppm which is attributed to an octahedrally coordinated aluminium centre.[45] The compound tris(di-iso-propylmalonato)aluminium, exhibiting a comparable coordination environment to the title compound showing a single ^{27}Al-NMR signal at 5.3 ppm.[44]

Thermogravimetric analysis of the aluminum precursor compound (Fig. 1a) under an oxygen and an argon atmosphere gives a ceramic yield of 8.15% close to the theoretically expected yield of 7.97%. However, under an argon atmosphere the obtained yield is 13.29% indicating an incomplete decomposition due to remaining organic residues under the reduced oxygen content. The included Gram–Schmidt-curve shows a maximum at 240 °C at which about 90% of the gaseous decomposition products are released under an oxygen atmosphere.

Differential scanning calorimetry (DSC) shows a corresponding exothermic peak at 230 °C (Fig. 1b). Additionally, mass spectrometry (Fig. 1c) and infrared spectroscopy (Fig. 1d), at the maximum of the Gram–Schmidt signal (240 °C), have been coupled with the thermogravimetric analysis, in order to investigate the gaseous decomposition products. The mass spectrum exhibits small, unspecific molecules like water, carbon monoxide, carbon dioxide, nitrogen monoxide, nitrogen dioxide

58

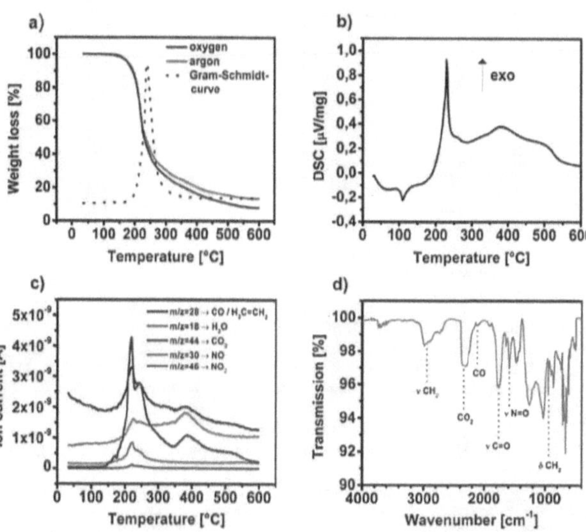

Fig. 1 (a) Thermogravimetric analysis of **1** in an oxygen atmosphere (blue) and an argon atmosphere (red), including the Gram–Schmidt curve (black dots), (b) differential scanning calorimetry (DSC) for **1** and (c) ionic current of different masses of the gaseous decomposition products of **1**. (d) Infrared spectrum of the gaseous decomposition products of **1** measured in an oxygen-atmosphere at the maximum of the Gram–Schmidt-signal (240 °C).

and ethylene. The infrared spectrum confirms the presence of carbon dioxide (2330 cm^{-1}), carbon monoxide ($\sim$2100 cm^{-1}) and nitrogen dioxide (1586 cm^{-1}). The vibrational bands at about 3000 cm^{-1} as well as 950 cm^{-1} can be assigned to ethylene. The vibrational band at 1760 cm^{-1}, most probably belongs to a carbonyl compound, however could not clearly be assigned to other potential candidates like ethanol, diethyl malonate, diethyl carbonate, diethyl oxalate or glycolic acid. Probably, there appear to be at least two carbonyl compounds present, resulting in overlaps of the vibrational bands. This is in contrast to the previously reported dimethyl malonato zinc complex.[46]

In order to investigate the material properties of the decomposition products, **1** was dissolved in 2-methoxethanol, spin

coated on glass and the obtained thin film of **1** annealed at different temperatures (200/250/300/350 °C). The X-ray diffraction (XRD) patterns of the solution processed thin films obtained at 200, 300 and 350 °C by decomposition of the precursor revealed an amorphous Al$_x$O$_y$ product throughout (Fig. 2a). This is highly desirable for a dielectric material where grain boundaries at crystalline interfaces can act as preferential paths for impurity diffusion and thus leakage currents, leading to an inferior performance of the desired dielectric.

The amorphous phase of Al$_x$O$_y$ was further confirmed by transmission electron microscopy (TEM, Fig. 2b) showing no crystalline domains on the nanometer scale throughout the sample. Nevertheless, and as expected at an annealing temperature of 800 °C, the Al$_x$O$_y$ material starts to crystallize forming α-Al$_2$O$_3$ (Fig. S11, ESI†).

The IR absorption bands in the range of 600–900 cm^{-1} are attributed to Al–O-vibrations of aluminium oxide (Fig. 3). The characteristic broad absorption bands at about 3300 cm^{-1} correspond to the hydroxyl groups attributable to Al(OH)$_3$ as well as to adsorbed water molecules which also show a deformation vibration mode at about 1590 cm^{-1}.[47] Additionally, there are absorption bands at about 1400 cm^{-1}. They originate due to the presence of carbon based adsorbents like C–H, C=C, C=O and possibly CO$_3$$^{2-}$ species due to the Lewis acidity of Al$_2$O$_3$.[48] In the case of samples Al$_x$O$_y$-300 and Al$_x$O$_y$-350 there are no other absorption bands present, which could be

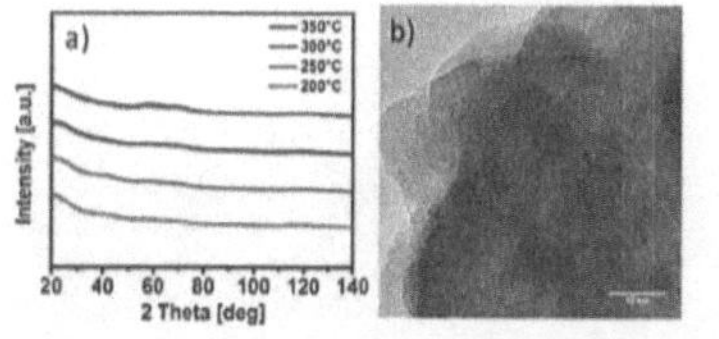

Fig. 2 (a) XRD patterns of solution processed Al$_x$O$_y$ synthesized at 200 °C, 250 °C, 300 °C and 350 °C. (b) TEM image of Al$_x$O$_y$ prepared at 350 °C.

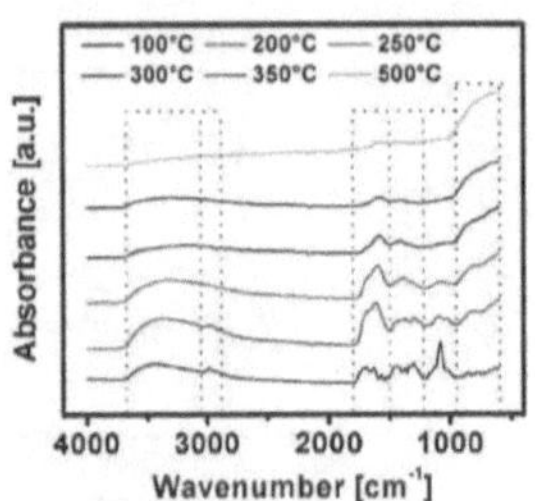

Fig. 3 IR spectra of Al-DEM-NO$_2$ annealed at various temperatures.

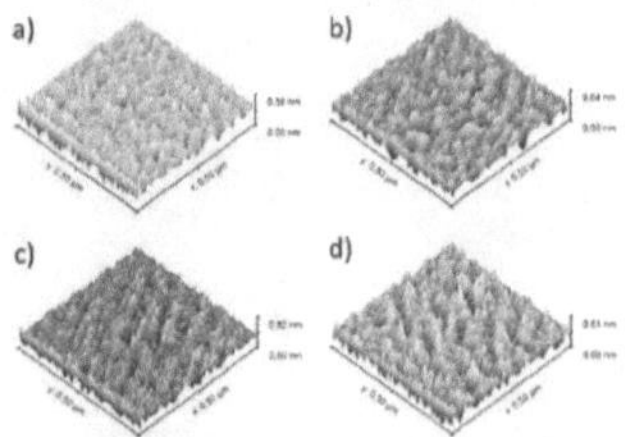

Fig. 4 AFM images of Al$_x$O$_y$ prepared at (a) 200 °C, (b) 250 °C, (c) 300 °C and (d) 350 °C.

attributed to organic residues. For the Al$_x$O$_y$-200 and Al$_x$O$_y$-250 samples there are still some absorption bands from residual ligand fragments. In the initial decomposition process starting from the precursor, first the absorption bands at about 1300–1600 cm^{-1} vanish.

These bands are attributable to NO$_2$ stretching vibrations. The Al$_x$O$_y$-200 sample still exhibits absorption bands at about 3000 cm^{-1} which are attributed to ν-CH$_2$ and ν-CH$_3$ stretching modes originating from the ethyl framework of the ligand. For the Al$_x$O$_y$-250 sample the ν-CH$_2$ and ν-CH$_3$ are not visible any longer. But still absorption bands are present at 1500–1800 cm^{-1} as well as at 1100 cm^{-1}, originating from –C–O and C=C vibrations. This indicates that these bonds from the ligand break at the latest at a temperature between or above 250 up to 300 °C. At an annealing temperature of 500 °C the hydroxide is fully converted into the amorphous oxide showing no remaining signals of any organic constituents.

Scheme 2 shows the thermal decomposition of the dimethyl malonate ligand of the precursor as analyzed by TG/MS and TG/IR analysis and IR spectroscopy. Sustainable with TG/MS and TG/IR analysis and IR-spectroscopy of the decomposition products, the first decomposition step is supposed to be the dissociation of the nitro group at about 230 °C. This step is followed by the dissociation of ethylene, leading to malonic acid as the remaining ligand fragment. The ethylene thermally decomposes in an oxygen atmosphere to carbon monoxide, carbon dioxide and water. The malonic acid, formed as a decomposition product, decomposes at higher temperatures as well to carbon monoxide, carbon dioxide and water.

Morphology and texture of the obtained films were studied by atomic force microscopy (AFM). Therefore, the precursor solution is spin-coated onto ITO-coated glass substrates and subsequently annealed at different temperatures. AFM (Fig. 4)

reveals a uniform and crack-free film formation with a very smooth surface. The roughness (R_{RMS}) of the surface for samples annealed at 200 °C, 250 °C, 300 °C and 350 °C is 0.33, 0.31, 0.35 and 0.30 nm, respectively. Additionally, XRR measurements have been performed in order to investigate the density and the thickness of the films. It is evident that the density increases with increasing annealing temperature. The Al$_x$O$_y$-200 and Al$_x$O$_y$-350 samples exhibit densities of 2.18 g cm^{-3} and 2.26 g cm^{-3}, respectively.

On the other hand, the thickness decreases with increasing annealing temperature. The thickness of the Al$_x$O$_y$-200 and Al$_x$O$_y$-350 samples were measured to be 170 nm and 122 nm, respectively and is in good agreement with spectroscopic ellipsometry measurements (Table S2 and Fig. S12 and S13, ESI†).

The Al$_x$O$_y$ films, deposited at various annealing temperatures (200, 250, 300 and 350 °C) in an ambient atmosphere, have been also investigated by X-ray photoelectron spectroscopy (XPS) to understand the chemical environment of the synthesized oxide. Fig. 5a–d exhibit the deconvoluted oxygen O1s core spectra of the prepared Al$_x$O$_y$ films. The O1s core spectra for all of the given annealing temperatures contain two peaks, at binding energies of 531.5 eV and 532.7 eV. The peak at 531.5 eV is attributed to oxygen associated with O^{2-} coordinated metal–oxygen bonds.[49,50] The peak at 532.7 eV is attributed to OH$^-$ associated to hydroxyl species.[51] The intensity of the two peaks at 532.7 eV and 531.5 eV change with increasing annealing temperature, whereby the peak associated to OH$^-$ decreases and the peak associated to O^{2-} increases (Fig. 5e). This is due to the decomposition of the precursor and conversion of the decomposition product of the precursor to form an aluminium oxide framework. This result is in good aggreement with the observed and improved dielectric performance for the capacitors fabricated from those films which were annealed consequently at higher temperatures.

Dielectric device properties of aluminium oxide

The dielectric properties of the solution processed Al$_x$O$_y$ films annealed at various temperatures have been measured using a metal–insulator–metal (MIM) structure (Fig. 6). The capacitors have been fabricated using ITO-coated glass substrates (140 nm, OLED-grade), with the ITO layer serving as the bottom electrode. With help of a shadow mask a 100 nm gold layer was deposited

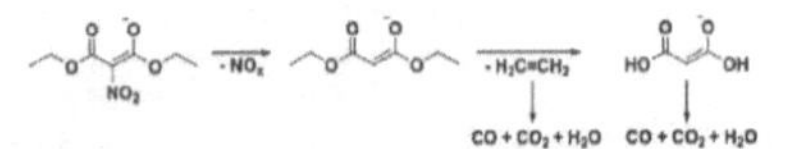

Scheme 2 Proposed thermal decomposition of the diethyl malonate ligand of 1 as analyzed by TG/MS and TG/IR analysis and IR-spectroscopy.

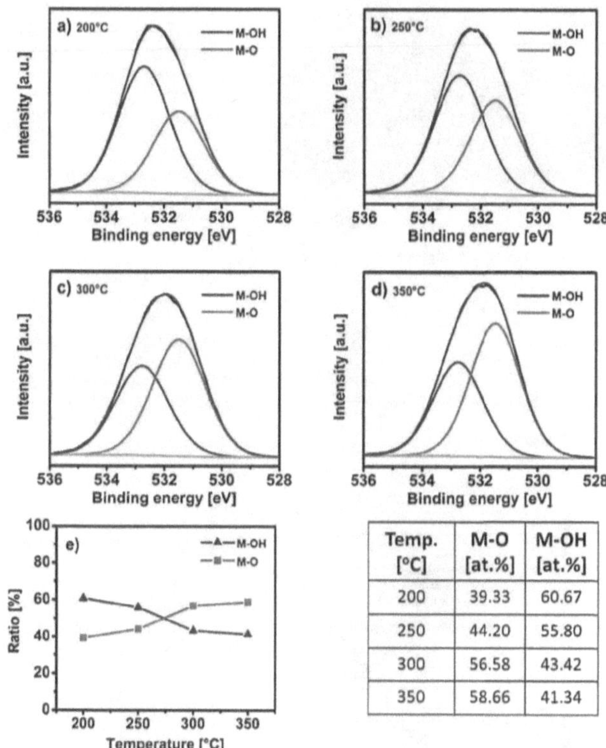

Fig. 5   Deconvoluted oxygen O1s core spectra of Al-DEM-NO₂ precursor annealed for 2 hours at (a) 200 °C; (b) 250 °C; (c) 300 °C; (d) 350 °C and (e) atomic concentrations of oxygen (M–O) and hydroxyl species (M–OH) derived from the deconvoluted XPS O1s spectra for the sample annealed at various temperatures.

at one side of the ITO layer, using the sputter technique. The deposited gold helps to enable the electrical contact with the ITO layer below, because aluminium oxide, which gets deposited on the top of the ITO layer, is too dense to get punctured through. The Al_xO_y dielectric layer was deposited by spin-coating and subsequent annealing at different temperatures, resulting in the film thicknesses of 122–170 nm. As the top electrodes 50 nm gold as well as a 10 nm interlayer titanium were sputtered on the Al_xO_y film, using another shadow mask. Fig. 7a shows the capacitance *vs.* frequency curves in the range of 10 Hz to 100 kHz. The Al_xO_y-350 sample exhibits an areal capacitance of 86 nF cm^{-2} at 10 kHz, while Al_xO_y-300, Al_xO_y-250 and Al_xO_y-200 exhibit areal capacities of 75 nF cm^{-2}, 51 nF cm^{-2} and 41 nF cm^{-2} respectively, at 10 kHz. The Al_xO_y-300 and Al_xO_y-350 samples show a high capacitance at

low frequencies which is attributed to the phenomenon known as capacitive reactance X_C. The X_C describes the imaginary part of the complex impedance of a capacitor $(Z = R + X_C)$, whereby the capacitance varies with respect to the frequency. It is described by the following equation: $X_C = 1/(2\pi fC)$. The measured capacities of these samples correspond to the calculated dielectric constants $k = 11.8$, 10.5, 7.6 and 6.5, respectively.

Additionally, breakdown measurements have been carried out (Fig. 7b). The measurements show that higher annealing temperatures correspond to higher breakdown voltages as well as less current leakage. The Al_xO_y-350 sample exhibits an electrical breakdown at 2.79 MV cm^{-1}. The electrical breakdown of Al_xO_y-300, Al_xO_y-250 and Al_xO_y-200 occur at 2.48 MV cm^{-1}, 2.16 MV cm^{-1} and 1.82 MV cm^{-1}, respectively. Although the breakdown voltage of these samples is lower than in previous

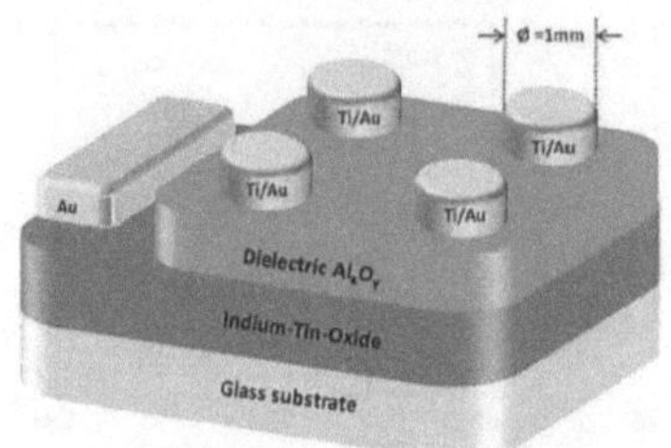

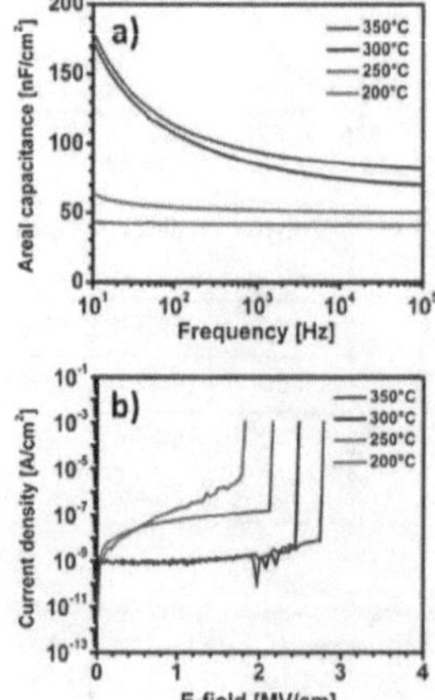

Fig. 6 Schematic illustration of the prepared capacitor. The grey layer represents the glass substrate, the purple layer represents a 140 nm ITO film serving as the bottom gate electrode and the blue layer illustrates the Al_xO_y dielectric film. The circular top electrodes are composed of 10 nm titanium and 50 nm gold and on the left-hand side is a 100 nm gold sacrificial contact.

Fig. 7 (a) Capacitance vs. frequency curves of solution processed Al_xO_y annealed at different temperatures. (b) Leakage current density versus electric field behavior of Al_xO_y-350, Al_xO_y-300, Al_xO_y-250 and Al_xO_y-200 samples.

reports, they still exhibit satisfying values. Therefore, the application of the Al_xO_y dielectric could be shown in a thin film transistor (TFT) device (see below).

The Al_xO_y-350 sample exhibits a very low leakage current density of 8.9×10^{-10} A cm^{-2} at 1 MV cm^{-1}. The samples Al_xO_y-300, Al_xO_y-250 and Al_xO_y-200 exhibit a leakage current density of 8.0×10^{-10}, 5.9×10^{-8} and 1.7×10^{-7} A cm^{-2}, respectively at 1 MV cm^{-1}. According to Wager et al. an ideal dielectric exhibits $J < 1 \times 10^{-8}$ A cm^{-2} at 1 MV cm^{-1},[54] This is indeed the case for Al_xO_y-350 as well as Al_xO_y-300.

It is especially noteworthy, that the atmosphere in which the measurements have been performed are crucial for the obtained dielectric properties. This is due to adsorbed water molecules from the ambient atmosphere which reside on the thin films and thus contribute to the parasitic resistance which then increases the capacitance drastically. The lower the chosen annealing temperature is, the higher this effect becomes, since there are more detrimental hydroxide groups residing on the surface enhancing the adsorption of additional water molecules. As a result, reports which have stated extraordinarily high capacitances even for very low annealing temperatures are a clear indicator that such studies have been performed under ambient conditions. Therefore, it becomes difficult to compare the dielectric properties measured in an ambient atmosphere with those measured under an inert atmosphere. Table 1 gives an overview of the studies performed so far. However, experimental details, especially measurement conditions, an ambient or an inert atmosphere, are sometimes not explicitly given in the cited work. It is obvious that the dielectric aluminium oxide Al_xO_y derived from the molecular precursor 1 shows a very good electrical performance compared to other aluminium based dielectrics. Only Yang and Avis reported slightly higher dielectric capacities for annealing temperatures in the range of 250–350 °C. However, with respect to the current leakage at 1 MV cm^{-1} both works do not fulfil the criterion of Wager et al. ($<10^{-8}$ A cm^{-2}) and are up to four magnitudes higher compared to the current leakage generated from the herein reported aluminium oxide precursor. This might be attributed to the difference in film thickness of these materials.

Finally, in order to demonstrate the feasibility of the electronic properties of the fabricated Al_xO_y dielectric it was applied in a TFT device. Therefore, a set of complete TFT

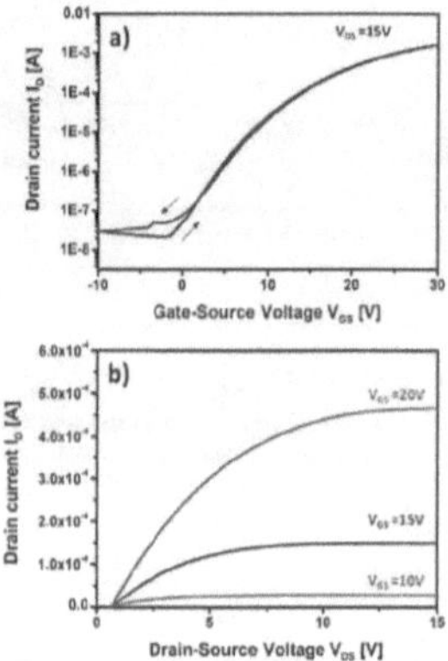

Fig. 8 Electrical characterization of TFT based on the Al_xO_y-350 and a solution processed indium zinc oxide (IZO) semiconductor processed at 350 °C. (a) Transfer characteristics of the device at V_{DS} = 15 V. (b) Output characteristics of the device measured at 10 V, 15 V and 20 V, respectively. (indicative arrows in the transfer characteristics showing the forward and reverse sweep directions).

Table 1 Overview of performance values of dielectric properties for solution processed Al_xO_y dielectric films

Ref.	Temp. (°C)	d (nm)	k	Areal cap. (nF cm^{-2}) (f)	Leakage [A cm^{-2}]	E_B (MV cm^{-1})
This work	350	122	11.8	86 (10 kHz)	8.9×10^{-10} (1 MV cm^{-1})	2.79
This work	300	125	10.5	75 (10 kHz)	8.0×10^{-10} (1 MV cm^{-1})	2.48
This work	250	132	7.6	51 (10 kHz)	5.9×10^{-8} (1 MV cm^{-1})	2.16
This work	200	170	6.5	41 (10 kHz)	1.7×10^{-7} (1 MV cm^{-1})	1.82
18	400	—	9.2	50 (120 Hz)	—	—
2	350	61	6.2	91 (1 MHz)	$<10^{-7}$ (1 MV cm^{-1})	>8
2	300	64	6.1	84 (1 MHz)	$\sim 10^{-7}$ (1 MV cm^{-1})	∼5
2	250	70	5.6	71 (1 MHz)	$>10^{-5}$ (1 MV cm^{-1})	∼6
5	300	70	6.3	80 (10 kHz)	6.3×10^{-5} (1 MV cm^{-1})	4
38	380	100	—	72 (10 kHz)	—	—
34	500	100	5.4	47.8 (1 MHz)	—	—
34	350	110	5.3	42.6 (1 MHz)	—	—
34	250	155	6.25	35.7 (1 MHz)	—	—
34	250	190	5.8	27 (1 MHz)	—	—
15	600	∼110	8.1	∼60* (1 MHz)	$>10^{-8}$ (1 MV cm^{-1})	∼2.5
15	500	∼110	6.6	∼49* (1 MHz)	$>10^{-8}$ (1 MV cm^{-1})	∼4.5
15	400	∼110	5.9	∼44* (1 MHz)	$>10^{-8}$ (1 MV cm^{-1})	∼2.5
15	300	∼110	—	—	$\sim 10^{-6}$ (1 MV cm^{-1})	∼1.5
52	600	162	10.0	56 (20 Hz)	—	2
35	250	100	—	47.7 (1 kHz)	$>10^{-6}$ (1 MV cm^{-1})	5
36	500	—	6.1	48.5 (1 MHz)	$>10^{-8}$ (1 MV cm^{-1})	—
53	400	—	—	∼50* (100 kHz)	3.41×10^{-8} (1 MV cm^{-1})	—

devices were fabricated by using the Al_xO_y-350 sample and a solution deposited amorphous indium zinc oxide (IZO) as the semiconductor material which was subsequently processed at 350 °C to its final state. The amorphous IZO film was generated by a solution technique using single source zinc and indium precursors as described by us previously.[55–57] A schematic representation of the TFT architecture is shown in the ESI† (Fig. S14). Electrical characterization of the TFTs was evaluated based on critical device performance metrics including the charge-carrier mobility (μ_{sat}) in the saturation regime, the threshold voltage (V_{th}) and the ratio of the current measured in the on-state and the off-state ($I_{on/off}$) of the final device. The measured transfer and output characteristics of the device are shown in Fig. 8. The measured devices exhibit a very good performance with a μ_{sat} of 7.1 cm^2 V^{-1} s^{-1}, a V_{th} of 8.7 V and a $I_{on/off}$ of 1.4×10^5, which clearly demonstrates the functionality by the effective implementation of the fabricated Al_xO_y dielectric when combined with an active IZO semiconductor in a TFT device.

Conclusions

In summary, we have demonstrated the synthesis and structural characterization of tris[[(diethyl-2-nitromalonato)]aluminium(III) **1** which presents a single source molecular precursor that decomposes at a fairly low transformation temperature into the metal oxide. Furthermore, the synthesized precursor is highly soluble (>20 wt%) in 2-methoxyethanol, a widely used solvent for the spin-coating technique. As a result this compound is suitable for the formation of aluminium oxide thin films, which represents a high k material with very good dielectric performance characteristics, and is thus a potential candidate for the replacement of SiO_2 as a gate dielectric. Furthermore, there is no need for additives or stabilizers using this solely molecular single source precursor approach for generating such thin films. The obtained very smooth Al_xO_y

films are amorphous for all annealing temperatures studied (200–350 °C) and possess a very low surface roughness of about 0.3 nm. In order to fabricate capacitors, ITO-coated glass served as a substrate as well as a gate electrode in processed capacitor devices. Therein the Al_xO_y layers were spin-coated on the ITO-coated glass and annealed at various temperatures (200, 250, 300 and 350 °C). Gold (50 nm) as well as a 10 nm interlayer of titanium was used as the top electrode. These capacitors exhibit satisfactory areal capacities even at moderate temperatures (86 nF cm^{-2} at 350 °C). Additionally, these capacitors exhibit very low current leakage of $<10^{-9}$ at 1 MV cm^{-1} (300 and 350 °C). Furthermore, implementation of the fabricated dielectric material with an IZO semiconductor processed at 350 °C demonstrates good TFT performance metrics with a μ_{sat} of 7.1 cm^2 V^{-1} s^{-1} and a $I_{on/off}$ of 1.4×10^5.

Conflicts of interest

The authors declare no conflict of interests.

Acknowledgements

TEM investigations were performed at ERC Jülich under contract with ERC-TUD1. We acknowledge Dr J. Engstler and S. Heinschke for performing TEM, XRD and ellipsometric studies and S. Foro for single crystal structure analysis of the title compound.

References

1 R. P. Ortiz, A. Facchetti and T. J. Marks, *Chem. Rev.*, 2010, **110**, 205–239.
2 W. Yang, K. Song, Y. Jung, S. Jeong and J. Moon, *J. Mater. Chem. C*, 2013, **1**, 4275.

3 B. N. Pal, B. M. Dhar, K. C. See and H. E. Katz, *Nat. Mater.*, 2009, **8**, 898.

4 G. Adamopoulos, S. Thomas, P. H. Wobkenberg, D. D. Bradley, M. A. McLachlan and T. D. Anthopoulos, *Adv. Mater.*, 2011, **23**, 1894–1898.

5 C. Avis and J. Jang, *J. Mater. Chem.*, 2011, **21**, 10649.

6 J. B. Kim, D. R. Kwon, K. Chakrabarti, C. Lee, K. Y. Oh and J. H. Lee, *J. Appl. Phys.*, 2002, **92**, 6739–6742.

7 X.-H. Zhang, S. P. Tiwari and B. Kippelen, *Org. Electron.*, 2009, **10**, 1133–1140.

8 P. Katiyar, C. Jin and R. J. Narayan, *Acta Mater.*, 2005, **53**, 2617–2622.

9 T. M. Klein, D. Niu, W. S. Epling, W. Li, D. M. Maher, C. C. Hobbs, R. I. Hegde, I. J. R. Baumvol and G. N. Parsons, *Appl. Phys. Lett.*, 1999, **75**, 4001–4003.

10 H. Q. Chiang, J. F. Wager, R. L. Hoffman, J. Jeong and D. A. Keszler, *Appl. Phys. Lett.*, 2005, **86**, 013503.

11 S. R. Thomas, P. Pattanasattayavong and T. D. Anthopoulos, *Chem. Soc. Rev.*, 2013, **42**, 6910–6923.

12 S. J. Kim, S. Yoon and H. J. Kim, *Jpn. J. Appl. Phys.*, 2014, **53**, 02BA02.

13 W. Xu, H. Wang, L. Ye and J. Xu, *J. Mater. Chem. C*, 2014, **2**, 5389.

14 P. K. Nayak, M. N. Hedhili, D. Cha and H. N. Alshareef, *Appl. Phys. Lett.*, 2013, **103**, 033518.

15 Y. Xu, X. Li, L. Zhu and J. Zhang, *Mater. Sci. Semicond. Process.*, 2016, **46**, 23–28.

16 H. Wang, W. Xu, S. Zhou, F. Xie, Y. Xiao, L. Ye, J. Chen and J. Xu, *J. Appl. Phys.*, 2015, **117**, 035703.

17 R. Branquinho, D. Salgueiro, L. Santos, P. Barquinha, L. Pereira, R. Martins and E. Fortunato, *ACS Appl. Mater. Interfaces*, 2014, **6**, 19592–19599.

18 G. Adamopoulos, S. Thomas, D. D. C. Bradley, M. A. McLachlan and T. D. Anthopoulos, *Appl. Phys. Lett.*, 2011, **98**, 123503.

19 W. Xu, M. Long, T. Zhang, L. Liang, H. Cao, D. Zhu and J.-B. Xu, *Ceram. Int.*, 2017, **43**, 6130–6137.

20 Q.-J. Sun, J. Peng, W.-H. Chen, X.-J. She, J. Liu, X. Gao, W.-L. Ma and S.-D. Wang, *Org. Electron.*, 2016, **34**, 118–123.

21 H. Wang, T. Sun, W. Xu, F. Xie, L. Ye, Y. Xiao, Y. Wang, J. Chen and J. Xu, *RSC Adv.*, 2014, **4**, 54729–54739.

22 Y. S. Rim, H. Chen, Y. Liu, S.-H. Bae, H. J. Kim and Y. Yang, *ACS Nano*, 2014, **8**, 9680–9686.

23 A. Liu, G. X. Liu, H. H. Zhu, F. Xu, E. Fortunato, R. Martins and F. K. Shan, *ACS Appl. Mater. Interfaces*, 2014, **6**, 17364–17369.

24 C.-G. Lee and A. Dodabalapur, *J. Electron. Mater.*, 2012, **41**, 895–898.

25 Y. B. Yoo, J. H. Park, K. H. Lee, H. W. Lee, K. M. Song, S. J. Lee and H. K. Baik, *J. Mater. Chem. C*, 2013, **1**, 1651.

26 J. H. Kim, B. D. Ahn, C. H. Lee, K. A. Jeon, H. S. Kang and S. Y. Lee, *Thin Solid Films*, 2008, **516**, 1529–1532.

27 A. Liu, G. Liu, H. Zhu, Y. Meng, H. Song, B. Shin, E. Fortunato, R. Martins and F. Shan, *Curr. Appl. Phys.*, 2015, **15**, S75–S81.

28 K. Song, W. Yang, Y. Jung, S. Jeong and J. Moon, *J. Mater. Chem.*, 2012, **22**, 21265.

29 J.-S. Park, J. K. Jeong, Y.-G. Mo and S. Kim, *Appl. Phys. Lett.*, 2009, **94**, 042105.

30 A. Liu, G. Liu, H. Zhu, B. Shin, E. Fortunato, R. Martins and F. Shan, *RSC Adv.*, 2015, **5**, 86606–86613.

31 W. Xu, H. Wang, F. Xie, J. Chen, H. Cao and J.-B. Xu, *ACS Appl. Mater. Interfaces*, 2015, **7**, 5803–5810.

32 H. Tan, G. Liu, A. Liu, B. Shin and F. Shan, *Ceram. Int.*, 2015, **41**, S349–S355.

33 E. Carlos, R. Branquinho, A. Kiazadeh, P. Barquinha, R. Martins and E. Fortunato, *ACS Appl. Mater. Interfaces*, 2016, **8**, 31100–31108.

34 H. Park, Y. Nam, J. Jin and B.-S. Bae, *RSC Adv.*, 2015, **5**, 102362–102366.

35 Y. S. Rim, H. Chen, T.-B. Song, S.-H. Bae and Y. Yang, *Chem. Mater.*, 2015, **27**, 5808–5812.

36 L. Zhu, Y. Gao, X. Li, X. Sun and J. Zhang, *J. Mater. Res.*, 2014, **29**, 1620–1625.

37 S. W. Smith, W. Wang, D. A. Keszler and J. F. Conley, *J. Vac. Sci. Technol., A*, 2014, **32**, 041501.

38 T. Lin, X. Li and J. Jang, *Appl. Phys. Lett.*, 2016, **108**, 233503.

39 A. H. Adl, P. Kar, S. Farsinezhad, H. Sharma and K. Shankar, *RSC Adv.*, 2015, **5**, 87007–87018.

40 Y. Zhang, G. Huang, L. Duan, G. Dong, D. Zhang and Y. Qiu, *Sci. China: Technol. Sci.*, 2016, **59**, 1407–1412.

41 D.-H. Lee, Y.-J. Chang, W. Stickle and C.-H. Chang, *Electrochem. Solid-State Lett.*, 2007, **10**, K51–K54.

42 Y. Zhao, G. Dong, L. Duan, J. Qiao, D. Zhang, L. Wang and Y. Qiu, *RSC Adv.*, 2012, **2**, 5307–5313.

43 B. Sykora, D. Wang and H. von Seggern, *Appl. Phys. Lett.*, 2016, **109**, 033501.

44 R. Lichtenberger, S. O. Baumann, M. Bendova, M. Puchberger and U. Schubert, *Monatshefte fur chemie*, 2010, **141**, 717–727.

45 M. Haouas, F. Taulelle and C. Martineau, *Prog. Nucl. Magn. Reson. Spectrosc.*, 2016, **94–95**, 11–36.

46 R. C. Hoffmann and J. J. Schneider, *Eur. J. Inorg. Chem.*, 2014, 2241–2247.

47 H. A. Al-Abadleh and V. H. Grassian, *Langmuir*, 2003, **19**, 341–347.

48 J. Gangwar, B. K. Gupta, P. Kumar, S. K. Tripathi and A. K. Srivastava, *Dalton Trans.*, 2014, **43**, 17034–17043.

49 V. Trouillet, H. Troesse, M. Bruns, E. Nold and R. White, *J. Vac. Sci. Technol., A*, 2007, **25**, 927.

50 B. R. Strohmeier, *Surf. Sci. Spectra*, 1994, **3**, 135–140.

51 C. Gao, X.-Y. Yu, R.-X. Xu, J.-H. Liu and X.-J. Huang, *ACS Appl. Mater. Interfaces*, 2012, **4**, 4672–4682.

52 D. X. Xia and J. B. Xu, *J. Phys. D: Appl. Phys.*, 2010, **43**, 442001.

53 J. Peng, C. Sheng, J. Shi, X. Li and J. Zhang, *J. Sol-Gel Sci. Technol.*, 2014, **71**, 458–463.

54 J. F. Wager, D. A. Keszler and R. E. Presley, *Transparent Electronics*, Springer, New York, 2008.

55 R. C. Hoffmann, M. Kaloumenos, S. Heinschke, E. Erdem, P. Jakes, R.-A. Eichel and J. J. Schneider, *J. Mater. Chem. C*, 2013, **1**, 2577–2584.

56 S. Sanctis, N. Koslowski, R. Hoffmann, C. Guhl, E. Erdem, S. Weber and J. R. J. Schneider, *ACS Appl. Mater. Interfaces*, 2017, **9**, 21328–21337.

57 S. Sanctis, R. C. Hoffmann, M. Bruns and J. J. Schneider, *Adv. Mater. Interfaces*, 2018, **5**, 1800324.

64

6.2 Synthesis, oxide formation, properties and thin film transistor properties of yttrium and aluminium oxide thin films employing a molecular-based precursor route

Yttrium oxide (Y_2O_3) is another promising dielectric material for the use as gate dielectric in TFT devices. Yttrium oxide possess a combination of favorable electrical properties like a high dielectric constant (14-18), wide band gap (5.8 eV), high breakdown voltage (> 3 MV cm^{-1}), high refractive index (1.9-2.0) and low dissipation factor (< 0.005).

This work is focusing on the combustion synthesis of yttrium oxide and aluminum oxide dielectric thin films, by introducing a single-source molecular precursor route. Therefore, a novel nitro functionalized malonato complex of yttrium (Y-DEM-NO$_2$ **1**) and well-defined urea nitrate complexes of yttrium (Y-UN **2**) and aluminum (Al-UN **3**) were utilized. All precursor compounds were studied by spectroscopic techniques (NMR/IR) as well as by single-crystal structure analysis for both urea nitrate coordination compounds. Differential scanning calorimetry (DSC) and thermogravimetry coupled with mass spectrometry and infrared spectroscopy (TG-MS/IR) was used to study the thermal decomposition of the precursors **1–3**, enabling a controlled thermal conversion of the precursors into dielectric thin films. The thermal transformation of the aluminum precursor **3** at temperatures between 200 and 350 °C, reveal amorphous products throughout, which is confirmed even on the nanometer scale by TEM investigations. Regarding the yttrium precursors, Y-DEM-NO$_2$ **1** exhibits first crystalline reflexes at 350 °C and forms cubic Y_2O_3 in the space group $Ia\bar{3}$ at 500 °C. The thermal transformation of the yttrium precursor **2** already exhibits crystalline reflexes at 250 °C, finally forming cubic Y_2O_3 in the space group $Ia\bar{3}$ at 600 °C. The Y_xO_y and Al_xO_y thin films, generated by precursors **1-3**, were integrated within capacitor devices, in order to investigate the dielectric properties. Capacitor devices based on Y_xO_y-(1) display dielectric behavior at PDA temperatures of 200-350 °C, with areal capacity values ranging from 21 up to 84 nF cm^{-2} at 10 kHz. The Y_xO_y-(2) thin films start to perform at temperatures T $\geq$ 300 °C, exhibiting capacity values of 111 and 131 nF cm^{-2} at 10 kHz for Y_xO_y-(2)-300 and Y_xO_y-(2)-350, respectively. Finally, the thermal processing of Al_xO_y dielectric thin films, derived from precursor **3**, exhibit remarkable high capacities per unit area of 184, 216 and 259 nF cm^{-2} at 10 kHz for Al_xO_y-250, Al_xO_y-300 and Al_xO_y-350 respectively. The current leakage density amounts in all cases (for T $\geq$ 300 °C) less than 1.0 x 10^{-9} A cm^{-2} at 1 MV cm^{-1} and the electrical breakdown occurs at electric fields E_{BD} > 2 MV cm^{-1}. The increased performance at elevated temperatures corresponds to the enhanced conversion of the intermediate hydroxy species into the respective metal oxide, determined by XPS. Finally, a solution-processed Y_xO_y based TFT was manufactured employing the precursor Y-DEM-NO$_2$. The device exhibits decent TFT characteristics with a saturation mobility (μ_{sat}) of 2.1 cm^2 V^{-1} s^{-1}, a threshold voltage (V_{th}) of 6.9 V and an on/off current ratio ($I_{on/off}$) of 7.6 x 10^5.

RSC Advances

PAPER

Cite this: *RSC Adv.*, 2019, **9**, 31386

Synthesis, oxide formation, properties and thin film transistor properties of yttrium and aluminium oxide thin films employing a molecular-based precursor route†

Nico Koslowski,[a] Rudolf C. Hoffmann,[a] Vanessa Trouillet,[b] Michael Bruns,[b] Sabine Foro[c] and Jörg J. Schneider [*a]

Combustion synthesis of dielectric yttrium oxide and aluminium oxide thin films is possible by introducing a molecular single-source precursor approach employing a newly designed nitro functionalized malonato complex of yttrium (Y-DEM-NO$_2$ **1**) as well as defined urea nitrate coordination compounds of yttrium (Y-UN **2**) and aluminium (Al-UN **3**). All new precursor compounds were extensively characterized by spectroscopic techniques (NMR/IR) as well as by single-crystal structure analysis for both urea nitrate coordination compounds. The thermal decomposition of the precursors **1–3** was studied by means of differential scanning calorimetry (DSC) and thermogravimetry coupled with mass spectrometry and infrared spectroscopy (TG-MS/IR). As a result, a controlled thermal conversion of the precursors into dielectric thin films could be achieved. These oxidic thin films integrated within capacitor devices are exhibiting excellent dielectric behaviour in the temperature range between 250 and 350 °C, with areal capacity values up to 250 nF cm^{-2}, leakage current densities below 1.0×10^{-9} A cm^{-2} (at 1 MV cm^{-1}) and breakdown voltages above 2 MV cm^{-1}. Thereby the increase in performance at higher temperatures can be attributed to the gradual conversion of the intermediate hydroxy species into the respective metal oxide which is confirmed by X-ray photoelectron spectroscopy (XPS). Finally, a solution-processed Y$_x$O$_y$ based TFT was fabricated employing the precursor Y-DEM-NO$_2$ **1**. The device exhibits decent TFT characteristics with a saturation mobility (μ_{sat}) of 2.1 cm^2 V^{-1} s^{-1}, a threshold voltage (V_{th}) of 6.9 V and an on/off current ratio ($I_{on/off}$) of 7.6×10^5.

Received 12th July 2019
Accepted 20th September 2019

DOI: 10.1039/c9ra05348d

rsc.li/rsc-advances

Introduction

High-k dielectrics based on metal oxides have gained remarkable attention due to their applicability for a variety of electronic and optoelectronic applications.[1] To date, most of the oxide thin-film transistors (TFTs) reported are based on the conventional dielectric SiO$_2$, which may exhibit a higher leakage current and hence require higher operational voltages to enhance the electronic properties of the TFTs.[2,3] Although significant progress has been achieved in terms of oxide semiconductors, investigations concerning the application of novel oxide dielectrics are still helpful to improve the current situation. Currently, various binary metal oxides like TiO$_2$,[4] Ta$_2$O$_5$,[5] HfO$_2$,[6] ZrO$_2$,[4,7–9] Al$_2$O$_3$[4,10–13] and Y$_2$O$_3$[4,14–18] have already shown promising results and with respect to their electronic performance, they are in the realm that they potentially can replace SiO$_2$ as gate dielectric in TFT devices.

However, these high-k materials are mostly deposited by expensive vacuum-based processes and create challenges in processing in order to further extend their application towards large-area applications.[19] Hence, a significant amount of research has been dedicated to the solution processing of amorphous metal oxides due to the possibility of cost-efficient, large-area deposition as well as the use of printing techniques to deposit these building blocks for electronic applications.[2,19,20] Among the various high-k dielectrics, solution-processed amorphous aluminium oxide, Al$_2$O$_3$, has already demonstrated its potential as excellent choice for the use as gate dielectric in TFTs.[4,21] Thereby, its impressive dielectric performance can be related to a minor accumulation of charge carriers within the bulk dielectric and a small trap density at the semiconductor/

[a] *Fachbereich Chemie, Eduard-Zintl-Institut für Anorganische und Physikalische Chemie, Technische Universität Darmstadt, Alarich-Weiss-Str. 12, 64287 Darmstadt, Germany*

[b] *Institute for Applied Materials (IAM-ESS), Karlsruhe Nano Micro Facility (KNMF), Karlsruhe Institute of Technology (KIT), Hermann-von-Helmholtz-Platz 1, 76344 Eggenstein-Leopoldshafen, Germany*

[c] *Department of Material Science, Technische Universität Darmstadt, Alarich-Weiss-Str. 8, 64287 Darmstadt, Germany*

† Electronic supplementary information (ESI) available. CCDC 1937519 for precursor **2**. For ESI and crystallographic data in CIF or other electronic format see DOI: 10.1039/c9ra05348d

dielectric interface.[4] Recently, amorphous aluminium oxide thin films with very good dielectric properties are also accessible by a molecular single-source precursor approach reported by our group.[22] We have devised a route to Al_xO_y dielectric, its synthesis and structural elucidation by employing the molecular coordination compound tris[(diethyl-2-nitromalonato)] aluminium(III) (Al-DEM-NO$_2$).

Yttrium oxide, Y_2O_3, is yet another promising high-k material due to a combination of favourable individual electrical properties like a wide bandgap (5.8 eV), a high refractive index (1.9–2.0), a high dielectric constant (14–18), a low dissipation factor (<0.005) and finally a high breakdown voltage (>3 MV cm^{-1}).[23] Furthermore, Y_2O_3 has the ability to interact intimately with the oxide semiconductor by forming a chemical surface bond leading to an improved interaction and formation of an intimate dielectric-semiconductor interface. As a result, low electron trap densities and thus low leakage currents can be realized.[24] Despite such advantages, only a few studies have investigated the potential towards its application in TFT technology. Adamopolous *et al.* executed a direct comparison between solution-processed Al_2O_3 and Y_2O_3 dielectrics, using aluminium acetylacetonate and yttrium acetylacetonate hydrate as precursors. Al_2O_3 and Y_2O_3 dielectric thin films were deposited by spray pyrolysis and subsequent calcined at 400 °C in air. As a result, electrical measurements performed in vacuum (10^{-5} mbar) revealed a significantly higher capacitance for the Y_2O_3 based dielectrics.[18] Liu *et al.* demonstrated the fabrication of Y_2O_3 dielectrics *via* an aqueous route, using yttrium nitrate hydrate as the precursor and deionized water as solvent. The thin films were deposited by means of spin-coating and subsequent calcination at 300 °C. The generated Y_2O_3 dielectric exhibited very low leakage current (10^{-9} A cm^{-2} at 1 MV cm^{-1}) and a dielectric constant of 14.8.[15]

In general, the majority of the studies use conventional metal salts like nitrates[14,15,17,25–27] or chlorides[16,12,13,28] as precursors, which possess some drawbacks like the additional requirement for additives or stabilizers, which exert an uncertainty on the film formation and thus the electrical properties of the final device. In the case of metal chlorides, acidic by-products like HCl or Cl$^-$-ion trace impurity present in the final ceramic can deteriorate the overall device performance.[29–31]

A meaningful strategy to circumvent such disadvantages is the use of molecular precursors, which typically start to decompose at moderate temperatures of about 200 °C. As a drawback, organic residues from the ligand framework remain in the dielectric thin films at low annealing temperatures. The complete decomposition of the ligand framework usually requires temperatures up to 500 °C. A common approach dealing with that issue uses the systematic introduction of reactive nitro or nitroso functionalities into the ligand framework, enhancing the exothermic nature of the decomposition of the precursor.[32,33] Herein we have chosen diethyl-2-nitromalonate as ligand in the yttrium oxide single-source precursor molecule. A common approach used is "combustion synthesis", which is based on a "fuel/oxidizer" reaction and enables a complete conversion of the precursor at much lower temperatures which are typically employed in decomposition

reactions. Often, urea or acetylacetone serve as "fuel" and metal nitrates as "oxidizer". Initiated by the thermal decomposition, redox reactions between the nitrate anion and the urea molecules occur, enhancing the conversion into the metal oxide and leading to various nitrogen compounds as by-products. Concerning solution-processed dielectrics this approach already showed good applicability. However, the detailed decomposition mechanism of the combustion synthesis has not been clarified in full so far. Besides that, aqueous solutions of metal nitrates and urea as additive usually require ageing of the solution over a more extended period of time to initiate the combustion effect. In order to gain more control and reproducibility in the combustion process, we have chosen a new approach combining the "oxidizer" (nitrate ions) and the "fuel" (urea ligands) in one defined coordination compound pre-defined in one molecule. This facilitates systematic studies of the respective metal oxide formation in a more accurate manner.

In this work we demonstrate the synthesis and structural elucidation of bis(diethyl-2-nitromalonato) nitrato yttrium(III) **1** as well as urea nitrate coordination compounds of yttrium(III) **2** and aluminium(III) **3** and their applicability as molecular single-source precursor for the formation of dielectric thin films of yttrium oxide and aluminium oxide, respectively. The molecular structures of the new oxide precursor compounds **2** and **3** were identified by single-crystal X-ray diffraction and spectroscopic techniques (IR, ^{1}H-, ^{13}C-, DEPT- and ^{27}Al-NMR). The oxide dielectric thin films were obtained by spin-coating of the respective molecular precursor, using either 2-methoxyethanol as solvent in the case of **1** and water for **2** and **3**. Subsequent calcination at moderate temperatures between 250 and 350 °C of these precursor molecules occur without requirement of any additive and yields dielectric thin films with excellent electrical performances. Precursor **1**, **2** and **3** are thus capable for the solution processing of aluminium and yttrium oxide and their subsequent use as gate dielectrics in TFTs.

Experimental section

Synthesis and characterization

Synthesis and characterization of bis(diethyl-2-nitromalonato) nitrato yttrium(III) (Y DEM-NO$_2$) 1. 1.825 g (5 mmol) yttrium nitrate pentahydrate was dissolved in 50 mL ethanol. 3.330 g (15 mmol) ammonium 2-nitro-diethyl malonate were added under stirring to the clear solution. Thereby a yellow colouring of the solution as well as a white precipitate was observable. The solution was stirred for 18 hours. After the filtration, the yellow solution was concentrated by rotary evaporation leaving a yellow oil. The oil was dissolved in 20 mL dichloromethane (DCM) and filtered through a 0.2 μm polytetrafluoroethylene (PTFE) syringe filter. Finally, 200 mL of *n*-pentane was added to the clear yellow solution forming a solid yellow product. The yellow product was finally dried under vacuum <10^{-4} bar giving 2.2 g (62.67%). Elemental analysis: C 30.76%, H 3.87% and N 7.90%, calc. for YC$_{14}$H$_{20}$O$_{15}$N$_3$ (559.22 g mol^{-1}) C 30.07%, H 3.60% and N 7.51%. ^{1}H-NMR (300 MHz, [d$_4$] methanol) $\delta = 1.30$ (t, –CH$_3$); 4.20 (q, –CH$_2$) ppm. ^{13}C-NMR (300

67

MHz, [d_6] dimethyl sulfoxide) 14.72 (–CH$_3$); 59.93 (–CH$_2$); 110.31 (–C–NO$_2$) 163.47 (–C=O) ppm.

IR = 2985 (s, ν_{CH}), 2941 (s, ν_{CH}), 1742 (s, $\nu_{C=O}$), 1638 (s, $\nu_{N=O}$), 1426 (s, $\nu_{as,N-O}$), 1323 (s, ν_{CN}), 1255 (s, $\nu_{s,N-O}$), 1067 (s, ν_{C-O}), 862 (s, $\delta_{O=N-O}$).

Synthesis and characterization of dinitrato tetra(urea) yttrium(III)-nitrate [Y(urea)$_4$(NO$_3$)$_2$](NO$_3$) (Y-UN) 2. 3.65 g (10 mmol) yttrium nitrate pentahydrate was dissolved in 50 mL butanol. 2.40 g (40 mmol) urea is added under stirring, whereby precipitation is observable after a few minutes. The dispersion is stirred for 3 hours. Afterwards the white powder is separated by centrifugation. The white powder is dissolved in methanol and crystallized *via* diffusion of diethyl ether into the methanol solution. After 48 hours white crystals are formed in the solvent mixture. After decantation of the solvent, the white crystals are dried under vacuum <10^{-4} bar giving a yield of 5.2 g (81.89%). Elemental analysis: C 9.43%, H 3.34% and N 29.05%; calc. for YC$_4$H$_{16}$O$_{13}$N$_{11}$ (515.14 g mol^{-1}) C 9.33%, H 3.13% and N 29.91%. ^{1}H NMR (500 MHz, [d_6] dimethyl sulfoxide) δ = 5.54 (s, –NH$_2$) ppm. ^{13}C{1H}^{34} NMR (500 MHz, [d_6] dimethyl sulfoxide) 160.43 (–C=O) ppm.

IR = 3472 (s, ν_{NH}), 3366 (s, ν_{NH}), 3212 (s, ν_{NH}), 1625 (s, ν_{CO}), 1578 (s, δ_{NH}), 1489 (s, ν_{NO}), 1468 (s, ν_{NO}), 1448 (s, ν_{NO}), 1397 (s, ν_{NO}), 1343 (s, ν_{CN}), 1294 (s, ν_{CN}), 1145 (m, δ_{NH}), 1030 (m, δ_{NH}), 819 (w, δ_{NH}), 772 (w, δ_{NH}), 745 (w, δ_{NH}), 584 (m, δ_{CN}), 527 cm^{-1} (m, δ_{CN}). The detailed crystallographic data for **2** is provided in the ESI.†

Synthesis and characterization of hexakis(urea) aluminium(III)-nitrate [Al(urea)$_6$](NO$_3$)$_3$ (Al-UN) 3. 7.21 g (120 mmol) urea was dissolved in 240 mL ethanol. 7.50 g (20 mmol) aluminium nitrate nonahydrate was added under stirring, whereby precipitation is observable after a few minutes. The dispersion was stirred for 3 hours. Afterwards, the white precipitate was separated by centrifugation. The white powder is dissolved in methanol and crystallized by addition of diethyl ether. After 48 hours white crystals are formed and dried after decantation of the solvent under vacuum < 10^{-4} bar. The yield was 7.71 g (67.24% of the theory). Elemental analysis: C 12.60%, H 4.21% and N 37.48%, calc. for AlC$_6$H$_{24}$O$_{15}$N$_{15}$ (573.33 g mol^{-1}) C 12.57%, H 4.22% and N 36.65%. ^{1}H-NMR (500 MHz, [d_6] dimethyl sulfoxide) δ = 5.48 (s, –NH$_2$) ppm. ^{13}C{1H} NMR (500 MHz, [d_6] dimethyl sulfoxide) 160.21 (–C=O) ppm. ^{27}Al NMR (500 MHz, [d_4] methanol) δ = −9.6 (Al–O) ppm.

IR = 3447 (s, $\nu_{NH,out\ of\ phase}$), 3339 (s, $\nu_{NH,in\ phase}$), 3234 (s, $\nu_{NH,in\ phase}$), 1628 (s, ν_{CO}), 1571 (s, δ_{NH}), 1501 (s, δ_{NH}), 1331 (s, ν_{CN}), 1153 (m, δ_{NH}), 1035 (m, δ_{NH}), 828 (w, δ_{NH}), 763 (w, δ_{NH}), 618 (m, δ_{CN}), 541 (m, δ_{CN}), 421 cm^{-1} (m, δ_{NH}). The detailed crystallographic data for **3** is provided in the ESI.†

Oxide formation from precursor molecules 1, 2 and 3 and capacitor device fabrication therefrom. The precursor solution of **1**, **2** or **3** were prepared by dissolving 20 wt% of Y-DEM-NO$_2$ **1** in 2-methoxyethanol and 20 wt% of Y-UN **2** and Al-UN **3** in deionized water, followed by subsequent sonication for 20 minutes and filtering the solution through a 0.2 mm polytetrafluoroethylene (PTFE) syringe filter for **1** and **3**. In the case of precursor **2**, the solution was vigorously stirred for 2 hours at 60 °C, prior to filtration. ITO-coated glass substrates (140 nm,

OLED-grade) were used for the fabrication of the capacitors. At first, the substrates were cleaned in deionized water, acetone and isopropanol, for 10 minutes each using an ultrasonic bath. In order to enable the electrical contact with the ITO layer, a 100 nm gold layer was sputter-deposited using a shadow mask. Prior to spin-coating, the substrates were exposed to an air-plasma for two minutes to enhance the hydrophilicity of the substrate. Subsequently the clear precursor solutions were spin-coated on the substrate and annealed at different temperatures (200–350 °C). The spin-coating parameters were (20 s; 3000 rpm) for **1** and (20 s; 4000 rpm) for **2** and **3**, respectively. This procedure was repeated three times for all precursors. The thickness of the layers of 156–286 nm for **1**, 82–87 nm for **2** and 59–74 nm for **3** was determined by spectroscopic ellipsometry. Gold top electrodes (50 nm) were sputtered on the Y$_x$O$_y$ films with the help of another shadow mask. In the case of the Al$_x$O$_y$ based capacitors a 50 nm gold layer, as well as a 10 nm interlayer titanium, served as top electrodes. Y$_x$O$_y$ thin films, using Y-DEM-NO$_2$ as precursor, were annealed at 200, 250, 300 and 350 °C and are abbreviated as Y$_x$O$_y$-(1)-200, Y$_x$O$_y$-(1)-250, Y$_x$O$_y$-(1)-300 and Y$_x$O$_y$-(1)-350, respectively. The Y$_x$O$_y$ thin films, using Y-UN as precursor, were annealed at 300 and 350 °C and are abbreviated as Y$_x$O$_y$-(2)-300 and Y$_x$O$_y$-(2)-350. The calcination of Al$_x$O$_y$ thin films occurred at 250, 300 and 350 °C and are abbreviated as Al$_x$O$_y$-250, Al$_x$O$_y$-300 and Al$_x$O$_y$-350, respectively.

Thin-film transistor fabrication

For the fabrication of the TFT device an indium zinc oxide (IZO) semiconductor was introduced by employing established oximato precursor compounds, previously reported by our group.[22,14–36] Solutions of the respective indium and zinc precursors were prepared by dissolving 1 wt% in 2-methoxyethanol (ratio In : Zn, 6 : 4), spin-coated at 2500 rpm for 20 seconds onto the Y$_x$O$_y$-(1)-350 dielectric and subsequently annealed at 350 °C. Performing of three iterations of the coating procedure results in a film thickness of ∼12 nm.

Finally, gold source-drain electrodes (W/L = 2 mm/100 mm) were sputter deposited onto the IZO semiconductor, using a shadow mask (2 rectangular areas of 2 mm × 1 mm, separated by a distance of 150 μm).

Materials characterization

NMR-spectroscopy was carried out at 500 MHz using a DRX500 spectrometer (Bruker BioSpin GmbH). IR-spectroscopy was carried out on a Nicolet 6700 (Thermo Fisher Scientific). The samples were measured in attenuated total reflection (ATR) without additional preparation. Thermogravimetric analysis (TGA) coupled with mass spectrometry (MS) and infrared (IR) spectroscopy was performed using TG 209N1 (Netzsch) coupled with Aelos QMS 403C (Netzsch) and a Nicolet iS10 spectrometer (Thermo Fisher Scientific). The samples were measured in an oxygen-atmosphere with a heating rate of 2.5 or 5 K min^{-1}, in the range of 50–800 °C, in a corundum crucible. Differential scanning calorimetry (DSC) was performed with STA 449 F3 Jupiter (Netzsch). X-ray diffraction (XRD) measurements were carried out using a MiniFlex 600 (Rigaku), using Cu-Kα-

radiation (600 W) in Bragg-Brentano geometry. Transmission electron microscopy (TEM) was carried out with an operating voltage of 200 kV, using a Tecnai F20 (FEI) system. The samples were prepared on lacey carbon-coated copper grids. Spectroscopic ellipsometry was carried out using a Woolham M-2000 V spectrometer (spectral range 370–1690 cm^{-1}) using the completeEASE software (version 6.29). Atomic force microscopy (AFM) was performed using MFP-3D (Asylum Research) equipped with silicon cantilevers. X-ray photoelectron spectroscopy (XPS) measurements were performed on a K-Alpha(+) XPS system (Thermo Fisher Scientific, East Grinstead, UK). The monochromated Al Kα X-ray source was used at a spot size of 400 mm. All spectra are referenced to the C 1s peak of hydrocarbons at 285.0 eV.

Electrical characterization

Impedance measurements were performed in a glovebox under inert conditions, using a ModuLab MTS System (Solartron Analytical Ltd) equipped with a probe station (Cascade Microtech, Inc). Impedance measurements were operated in the frequency range of 10 Hz to 100 kHz with amplitudes of 500 mA. The measurements of the breakdown voltage were carried out on a B1500A semiconductor device analyzer (Agilent).

TFT transfer and output characteristics were determined using a HP 4155A semiconductor parameter analyzer (Agilent) in a glove box under inert conditions.

Results and discussion

Synthesis of precursor molecules and materials characterization

The coordination compounds $[Y(urea)_4(NO_3)_2][NO_3]$ **2** and $[Al(urea)_6][NO_3]_3$ **3** were synthesized by the reaction of the

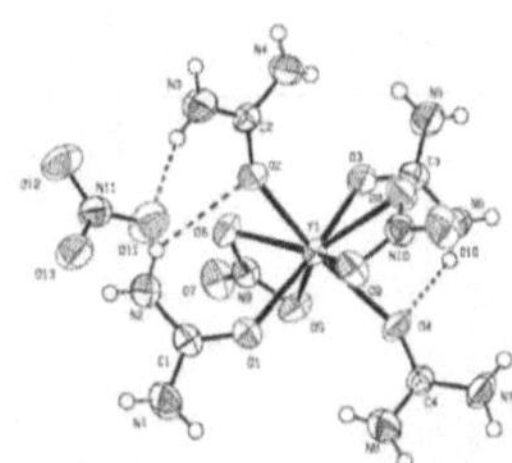

Fig. 2 ORTEP plot of the molecular structure of Y-UN **2**. Vibrational ellipsoids are drawn at the 50% probability. O–Y bond length are in the range 222–226 pm for the urea ligand and 243–249 pm for the nitrate ligand; O–Y–O bond angles are 90 ± 5° for the urea ligand.

respective metal nitrate hydrates with stoichiometric amounts of urea, using *n*-butanol for **2** and ethanol for **3** as solvent (Fig. 1). Compound **2** crystallizes in the space group *P*1̄ (Fig. 2) and **3** crystallizes in the space group *P*2₁/*n*. In both cases the neutrally charged urea molecules act as monodentate ligands and coordinate to the central metal atom by its oxygen atom in accord with Pearson's hard/soft acid base concept. For compound **2** the central metal is coordinated by four urea and two nitrate molecules, whereby the nitrate molecules act as bidentate ligands and coordinate to the yttrium centre by two oxygen atoms, leading to a double capped trigonal prism with coordination number eight for Y^{3+}. Although crystal quality of **3** so far precluded a satisfactory crystal structure refinement, a preliminary refinement allows to determine the connectivity giving an octahedral coordination (Fig. S1 ESI†). Additionally, the octahedral coordination environment of the aluminium centre in **3** is confirmed by ^{27}Al-NMR spectroscopy (Fig. S10†) giving a singlet peak at −9.6 ppm which is attributed to the octahedral coordination of the aluminium centre. The compound hexakis(urea) aluminium(ɪɪɪ)-chloride, exhibits a comparable coordination environment to precursor **3** displaying a singlet peak at −7.6 ppm in the ^{27}Al-NMR spectrum.[17] Although synthesis of the aluminium compound **3** has been reported,[17,38] its versatility with respect to an application was not demonstrated so far. Hexakis(urea) nitrate compounds with trivalent metal cations are also known for iron,[38] indium and gallium,[39] the latter two have been reported by our group, recently. $[Metal(urea)_4(NO_3)_2][NO_3]$ compounds as in the case of **2** are not known so far. A comparable compound exhibits three monodentate tetramethylurea ligands and three bidentate nitrate anions. Thereby the coordination polyhedron is a tricapped trigonal prism with coordination number nine for the yttrium central cation.[40]

It is noteworthy that the nitrate counter anion is crucial for the strong thermal decomposition of these complexes. This is due to the fuel/oxidizer reaction between the urea molecules and the nitrate species, resulting in an effective conversion of the ligand sphere and the formation of the respective metal oxides at moderate combustion temperatures.

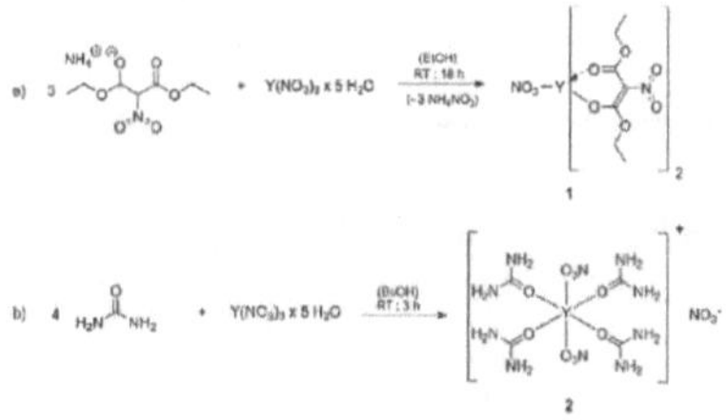

Fig. 1 (a) Schematic illustration of the synthesis of bis(diethyl-2-nitromalonato) nitrato yttrium(ɪɪɪ) (Y-DEM-NO₂) **1** and (b) and (c) showing the reaction scheme for the formation of the metal urea compounds of yttrium **2** and aluminium **3**.

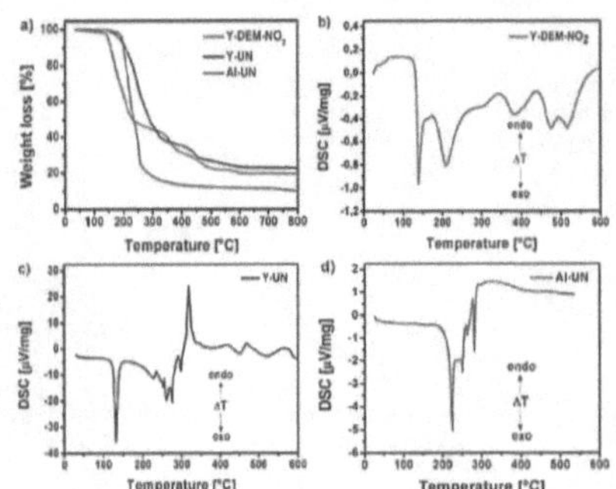

Fig. 3 (a) Thermo gravimetric analysis of 1, 2 and 3 in an oxygen atmosphere and (b–d) differential scanning calorimetry (DSC) for 1, 2 and 3.

The thermal decomposition of 1, 2 and 3 was carried out in an oxygen atmosphere to get an insight in the decomposition by-products of the precursors. The decomposition behavior of the yttrium precursors proceeded as a multi-step decomposition whereby significant mass loss occurs in the first step itself. The aluminium precursor proceeds as a one-step decomposition. The residual mass of all the precursors are in good agreement with the expected ceramic yield from the decomposition of the precursors in oxygen 1: $CY_{calc.}$ 20.19%, $CY_{meas.}$ 19.49%; 2: $CY_{calc.}$ 21.92%, $CY_{meas.}$ 22.73% and 3: $CY_{calc.}$ 8.90%, $CY_{meas.}$ 10.10%.

Differential scanning calorimetry (DSC) for 1 shows exothermic peaks at about 170, 235, 400 and 500 °C (Fig. 3b). The DSC for 2 displays a sharp exothermic peak between 100–150 °C,

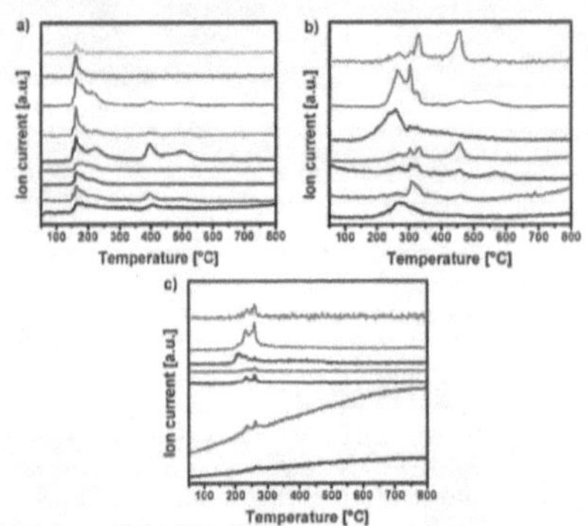

Fig. 4 MS intensities of (a) Y-DEM-NO₂ 1, (b) Y-UN 2 and (c) Al-UN 3 for m/z^+ peaks corresponding to the TG curves in Fig. 3 respectively.

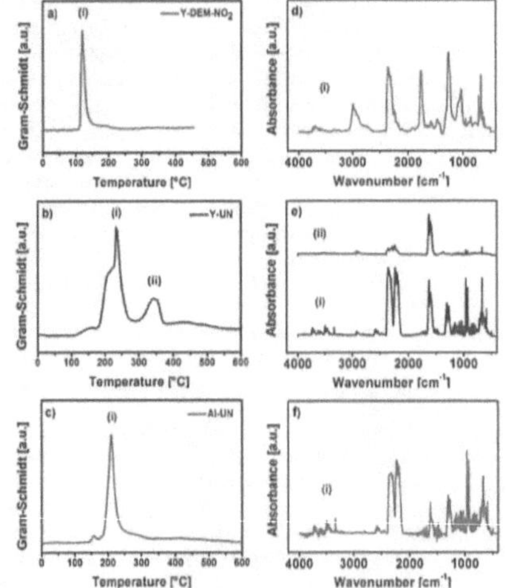

Fig. 5 Gram–Schmidt intensities (a–c) of 1, 2 and 3 and the corresponding IR signal intensities (d–f) according to the Gram–Schmidt signals of the precursors, respectively.

followed by less intense peaks between 250–300 °C. In the case of 3 the DSC shows a sharp exothermic peak at 225 °C, followed by less intense peaks between 250–300 °C. Furthermore, the differential scanning calorimetry of precursor 2 and 3 indicate a strong exothermic decomposition, which corresponds to the expected fuel/oxidizer reaction between the urea molecules and the nitrate anions. In order to investigate the gaseous decomposition products a thermogravimetric analysis with *in situ* mass spectrometry (Fig. 4) and infrared spectroscopy (Fig. 5) detection at the maximum of the Gram–Schmidt signal (120 °C for 1, 230 and 340 °C for 2 and 210 °C for 3) was performed.

Regarding 1 the detected gases could be assigned to water (m/z^+ 18), carbon monoxide (m/z^+ 28), nitric oxide (m/z^+ 30), carbon dioxide (m/z^+ 44) and nitrogen dioxide or ethanol (m/z^+ 46). But also larger fragments of the ligand can be found like C_2H_5O (m/z^+ 45), $C_3H_5O_2$ (m/z^+ 63) and $C_3H_7O_3$ (m/z^+ 91). The TG/IR at the maximum of the Gram–Schmidt signal confirms the release of carbon dioxide (CO_2: 2309 and 3356 cm⁻¹), as well as fragments containing alkyl groups (v_{CH} 2987 and 2939 cm⁻¹), and carbonyl groups ($v_{C=O}$ 1756 cm⁻¹).

Concerning precursors 2 and 3, the detected gaseous decomposition by-products are identical. Thereby, the releasing gases during the combustion synthesis could be assigned to ammonia (m/z^+ 17), water (m/z^+ 18), carbon monoxide (m/z^+ 28), nitric oxide (m/z^+ 30), isocyanic acid (m/z^+ 43), carbon dioxide (m/z^+ 44) and nitrogen dioxide (m/z^+ 46).

The corresponding IR signals detected at the maximum of the Gram–Schmidt signal clearly confirm for both precursors

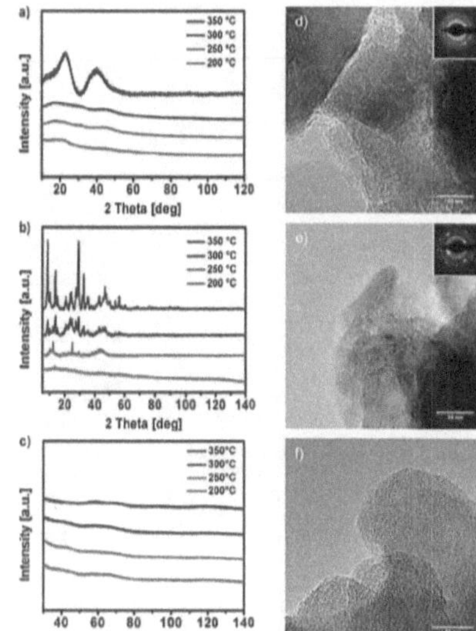

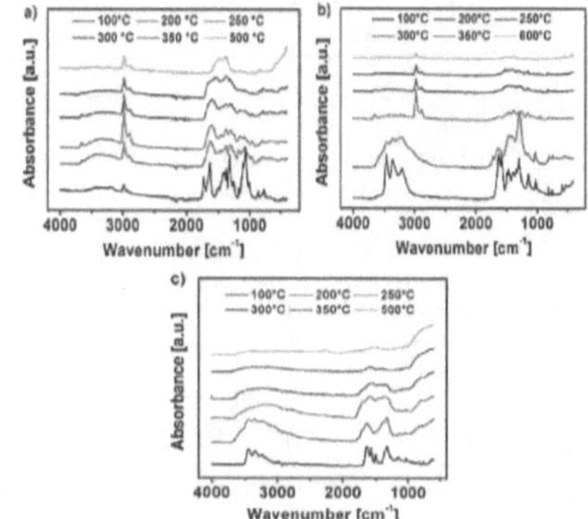

Fig. 7 (a–c) IR spectra of the thermal transformation of the precursors 1, 2 and 3 into the respective metal oxides at various temperatures.

Fig. 6 (a) and (b) XRD patterns of solution processed Y_xO_y from precursor 1 and 2 annealed at 200 °C, 250 °C, 300 °C and 350 °C. (c) XRD patterns of solution processed Al_xO_y from precursor 3 annealed at 200 °C, 250 °C, 300 °C and 350 °C. (d–f) TEM images of Y_xO_y and Al_xO_y from precursor 1, 2 and 3 prepared at 350 °C.

the evolution of ammonia (NH_3: 965 cm^{-1}), isocyanic acid ($HNCO$: 2239 cm^{-1}), carbon dioxide (CO_2: 2310 and 2359 cm^{-1}) and nitric acid (HNO_3: 1629 and 1305 cm^{-1}), which were detected at various decomposition stages in the temperature range between 200 °C and 350 °C. Nevertheless, a detailed mechanism for the thermal decomposition of the urea nitrate precursors remains challenging and is not clarified so far.

X-ray analysis of the solution-processed decomposition products of 1, 2 and 3 obtained at 200, 250, 300 and 350 °C (Fig. 6a, b and c) reveals a stepwise formation of the respective metal oxides. Thermal decomposition of 1 in the temperature range between 200–300 °C yields amorphous decomposition products, exhibiting first crystalline reflexes at 350 °C. After further annealing to 500 °C the material starts to crystallize, forming cubic Y_2O_3 in the space group $Ia3$ (Fig. S12a ESI†). In case of 2 the thermal transformation already exhibits crystalline reflexes at 250 °C, finally forming cubic Y_2O_3 in the space group $Ia3$ at 600 °C (Fig. S12b ESI†). Thermal transformation of 3 in the temperature range between 200–350 °C (200, 250, 300 and 350 °C) reveal an amorphous Al_xO_y product throughout. Furthermore, the amorphous phase of Al_xO_y was confirmed by transmission electron microscopy, which revealed no crystalline domains on the nanometer scale (TEM, Fig. 6f). However, at

an annealing temperature of 600 °C the Al_xO_y material starts to crystallize forming α-Al_2O_3 (Fig. S12c ESI†). The relatively low temperature of 600 °C for the formation of α-Al_2O_3, might be due to the strong exothermic fuel/oxidizer reaction between the urea molecules and the nitrate species during the decomposition of the precursor, resulting in local hot spots with higher temperatures.

The IR spectrum of the decomposition products of Y-DEM-NO_2 1 and Y-UN 2 is shown in Fig. 7a and b. The sharp IR absorption bands in the range of 2900–3100 cm^{-1} are attributable to Y–O-vibrations of yttrium oxide. The characteristic broad absorption bands at about 3400 cm^{-1} are associated with the hydroxyl groups attributable to $Y(OH)_3$ as well as to adsorbed water molecules which exhibit a deformation vibration mode at about 1590 cm^{-1}.[41]

The absorption bands at about 1400 cm^{-1} probably originate due to the presence of carbon-based adsorbents like C–H, C=C, C=O and possibly CO_3^{-2}-species due to the Lewis acidity of Y_2O_3.[42] In the case of the samples Y_xO_y-(1)-350, Y_xO_y-(2)-350, Y_xO_y-(2)-300 no other absorption bands are present, which could be attributed to organic residues.

At temperatures below 300 °C the samples exhibit some absorption bands from residual ligand fragments. Regarding precursor 2, the Y_xO_y-(2)-200 sample exhibits NH stretching modes in the range of 3100–3500 cm^{-1}, which originate from the amine groups of the urea ligands. For the Y_xO_y-(2)-250 sample, these vibrational bands are not visible any longer.

Regarding the initial decomposition process of precursor 1, absorption bands in the range of 1300–1700 cm^{-1} vanish at first. These bands are attributable to NO_2 stretching vibrations. For 1 absorption bands at about 3000 cm^{-1} are additionally expected for low annealing temperatures like 200 or 250 °C, which are attributed to ν-CH_2 and ν-CH_3 stretching modes originating from the ethyl framework of the 2-nitro-diethyl malonate ligand.[22] Unfortunately, the Y–O-vibrations of yttrium oxide are located in the same range, resulting in overlaps of the vibrational bands. At an annealing temperature of 500 °C for precursor 1 and 600 °C in the case of precursor 2, the hydroxide is fully converted into yttrium oxide, showing no remaining signals of any organic constituents.

Fig. 7c shows the IR spectrum of the decomposition products of Al-UN 3. In the range of 3100–3400 cm^{-1} are characteristic broad absorption bands present, originating from hydroxyl groups attributable to Al(OH)$_3$, as well as to adsorbed water molecules exhibiting a deformation vibration mode at about 1590 cm^{-1}.[11] Due to the Lewis acidity of Al$_2$O$_3$ the absorption bands at about 1400 cm^{-1} can also be assigned to the presence of carbon-based adsorbents like C–H, C=C, C=O and possibly CO$_3^{-2}$-species, similar as observed for the yttrium precursors 1 and 2. The absorption bands in the range of 400–900 cm^{-1} can be attributed to Al–O-vibrations of aluminium oxide.

In the case of the samples Al$_x$O$_y$-350 and Al$_x$O$_y$-300, no additional absorption bands attributable to organic residues are present. At annealing temperatures of 200 and 250 °C the samples display some additional absorption bands from residual ligand fragments. For the Al$_x$O$_y$-200 sample NH stretching modes in the range of 3100–3500 cm^{-1} are present, originating from the amine groups of the urea ligands, similar as observed for 2. At an annealing temperature of 250 °C the NH stretching modes are vanished. Finally, at an annealing temperature of 500 °C, the hydroxide is fully converted into aluminium oxide. Additionally, morphology and texture of the obtained thin films, prepared at 350 °C, were investigated by atomic force microscopy (AFM). AFM images of Y$_x$O$_y$-(1)-350 and Al$_x$O$_y$-350 reveal a uniform, smooth and crack-free film formation, while polycrystalline Y$_x$O$_y$-(2)-350 samples exhibit a relatively rough surface. The roughness (R_{RMS}) of the surface for the samples Y$_x$O$_y$-(1)-350, Y$_x$O$_y$-(2)-350 and Al$_x$O$_y$-350 is 0.31, 6.82, and 0.19 nm, respectively (Fig. S17a–c ESI†). We attribute the unexpected surface roughness of Y$_x$O$_y$-(2)-350 to local hot spots which are generated during the combustion synthesis of this precursor. Besides, a high degree of polycrystallinity of the sample may play a role. Indeed, comparable rough surfaces for yttrium oxide thin films have also been reported, exhibiting (R_{RMS}) of ~6.0 (ref. 14) and 17.40 nm (ref. 4) respectively.

XPS surface chemical analysis

Thin films obtained by thermal transformation of 1, 2 and 3 into the respective oxides Y$_x$O$_y$ and Al$_x$O$_y$ at temperatures of 200, 250, 300 and 350 °C were studied using X-ray photoelectron spectroscopy (XPS, Fig. 8a, 9a and 10a).

The O 1s core spectra of 1 (Fig. 8a) can be deconvoluted in three peaks with binding energies at 529.5, 532.0 and 534.3 eV. The peak at 529.5 eV corresponds to surface O^{2-} species coordinated to Y^{3+}.[43–46] Due to the fact that hydroxyl and carbonate species appear in the same range of binding energies[17] the peak at 532.0 eV can be attributed to both, OH$^-$ (ref. 44, 45) as well as CO$_3^{2-}$ species. The presence of such CO$_3^{2-}$ species coordinated to Y^{3+} is further supported by a characteristic peak at a binding

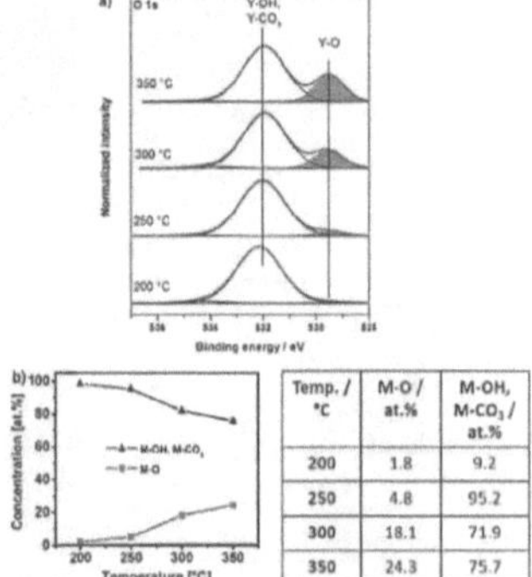

Fig. 8 (a) O 1s XPS core spectra of samples obtained from Y-DEM-NO$_2$ precursor 1 annealed for 2 hours each at 200 °C, 250 °C, 300 °C and 350 °C. (b) Atomic concentrations of oxygen (M–O) and hydroxyl as well as carbonate species (M–OH, M–CO$_3$), related to the total oxygen content and derived from the O 1s XPS spectra of 1 annealed at various temperatures.

energy of 290.3 eV in the C 1s core spectra[43] (Fig. S14a, ESI†). The very weak and only minor signal at 534.3 eV can be attributed to chemisorbed OH groups.[18]

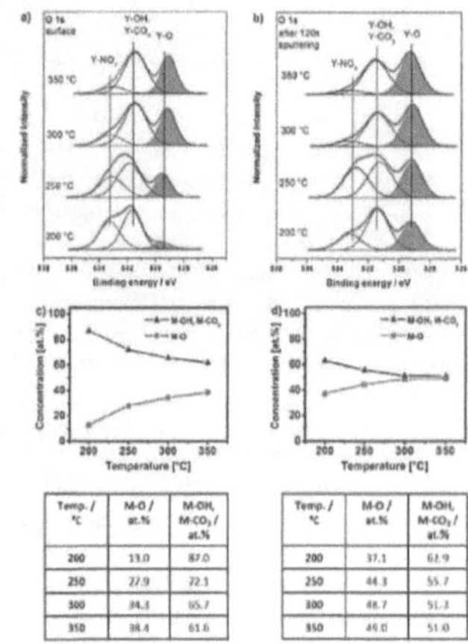

Fig. 9 O 1s XPS core spectra of samples obtained from Y-UN precursor 2 annealed for 2 hours at 200, 250, 300 and 350 °C (a) and (b) after 120 s sputtering (with cluster of 300 atoms with 8 keV energy). (c) Atomic concentrations of oxygen (M–O) and hydroxyl as well as carbonate species (M–OH, M–CO$_3$), related to the total oxygen content and derived from the O 1s XPS spectra taken from the surface as well as (d) in the sub surface layers close to the bulk and annealed at various temperatures.

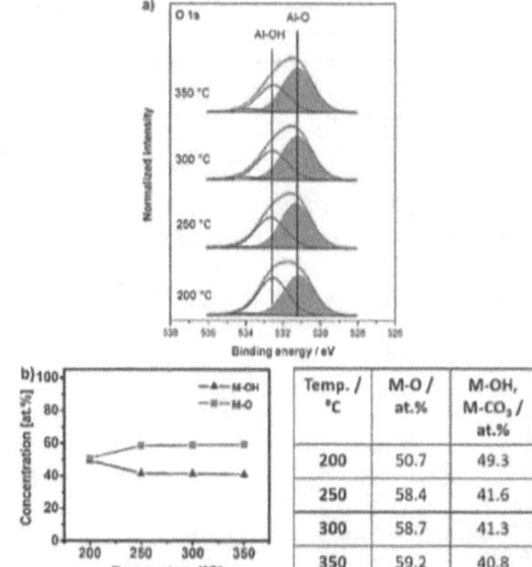

Temp. / °C	M-O / at.%	M-OH, M-CO₃ / at.%
200	50.7	49.3
250	58.4	41.6
300	58.7	41.3
350	59.2	40.8

Fig. 10 (a) O 1s XPS core spectra of samples obtained from Al-UN precursor 3 annealed for 2 hours each at 200 °C, 250 °C, 300 °C and 350 °C. (b) Atomic concentrations of oxygen (M–O) and hydroxyl species (M–OH), related to the total oxygen content and derived from the O 1s XPS spectra of 3 annealed at various temperatures.

Interestingly, for lower annealing temperatures of 200 and 250 °C minor amounts of NO_3^- species are also detectable and indicated by a peak at 407.4 eV in the N 1s core spectra[49,50] (Fig. S14b, ESI†). The presence of such NO_3^- species at these low annealing temperatures is in accord with a beginning thermal transformation process of precursor bis(diethyl-2-nitromalonato)nitrato yttrium(III) Y-DEM-NO₂ 1 into Y_xO_y. At an annealing temperature of 200 °C, O^{2-} coordinated metal-oxygen species are detected. However hydroxyl and carbonate species present the main species at that low temperature. Significant conversion into the final oxide starts in the temperature range between 250 and 300 °C. Consequently, the intensity of the Y–OH and Y–CO₃ fragments observed at 532.0 eV becomes smaller with increasing annealing temperature. In contrast, the peak associated with the Y–O species at 529.5 eV subsequently increases with increasing temperature (Fig. 8a), indicating a progressing transformation of the precursor molecule 1 and its conversion into the yttrium oxide framework. This finding corresponds nicely with the observation from the thermogravimetric analysis, whereby a significant mass loss occurs in the first decomposition step of precursor 1, ending at a temperature between 250 and 300 °C when the transformation to the oxide has already progressed significantly (Fig. 3a).

For the molecular yttrium oxide precursor Y-UN 2, the O 1s core spectra (Fig. 9a) as well as the spectra after 120 s of surface sputtering (Fig. 9b) have been recorded. In both cases, the O 1s core spectra were fitted according to three individual signals at binding energies of 529.0, 531.2 and 532.9 eV. The peak at

529.0 eV is in accord with O^{2-} species and the peak at 532.0 eV can be attributed to OH^- as well to CO_3^{2-} species both not discernable individually due to the narrow energy overlap of the two signals.[47] Furthermore, it becomes evident that the Y–O/(Y–OH, Y–CO₃) ratio is higher within the bulk of the material compared to its topmost surface composition (Fig. 9c–d). The peak at a binding energy of 532.9 eV can be attributed to NO_3^- species,[50] originating from the remaining precursor. The presence of such NO_3^- species on the surface as well as in the bulk material is further supported by an additional peak at a binding energy of 407.3 eV in the N 1s core spectra,[49] which is observed for all annealing temperatures studied (200–350 °C) (Fig. S16a and b, ESI†). The concentration of the NO_3^- species is higher on the surface than within the bulk material. Furthermore, the amount of NO_3^- species, evidenced in N 1s as well as in O 1s, is decreasing for both contributions, surface and subsurface when the annealing temperature increases which is in accord with the progressing conversion from 2 into the yttrium oxide framework.

At low annealing temperatures of 200 and 250 °C, when the decomposition is still incomplete, significant concentrations of up to 7.0 at% of NO_3^- species are detectable. Furthermore, the respective Y_xO_y-(2)-200 sample exhibits NH stretching modes in the range of 3100–3500 cm⁻¹ (Fig. 7b), originating from the still present residual amine groups of the urea ligands. This corroborates with the presence of a peak at 399.2 eV characteristic of nitrogen in amine groups (Fig. S16a and b, ESI†). This fact also supports a still incomplete transformation process of the ionic precursor 2 at that low annealing temperature. Such ionic residues in the developing oxide film can act as preferential parasitic pathways for the electrical current in an oxide dielectric. As a result, capacitor devices processed from materials derived from precursor 2 at such significant low temperatures of 200 and 250 °C exhibit electrical short-circuiting under voltage impact. At temperatures above 250 °C exothermic signals in the DSC analysis of 2, however prove the further ongoing extrusion of such ligand species during ongoing film formation in the thermal processing (Fig. 3c). At annealing temperatures of 300 and 350 °C, the nitrate concentration within the material is thus consequently reduced to 2.4 and 1.5 at%, respectively. This already results in a polycrystalline yttrium oxide based capacitor device which start to perform under these conditions (see upcoming section). Besides, it is remarkable that the generation of even thicker polycrystalline yttrium oxide dielectric films generated from precursor 2 at temperatures of 300 °C or above can further reduce the electrical short-circuiting in such metal oxide capacitor devices. This can be explained by a smaller statistical probability in generating parasitic electrical pathways from one electrode to the other, mediated by still remaining ionic ligand species in the film.

For the aluminium oxide precursor, Al-UN 3 the O 1s core spectra contain three peaks, at binding energies of 531.2 eV (M–O),[51,52] 532.6 eV (M–OH)[53] and 534.1 eV for all annealing temperatures studied (Fig. 10a). The broad and again weak minor signal at 534.1 eV can be attributed to chemisorbed OH groups.[48,54] The peak associated to the M–O functionality again subsequently increases and the peak associated to M–OH correspondingly decreases with increasing annealing temperature (Fig. 10b) similar as it is observed for the yttrium oxide precursors **1** and **2**. At an annealing temperature of 200 °C small amounts of NO_3^- species are present, indicated by a peak at a binding energy of 407.4 eV in the N 1s core spectra. Besides, the IR spectrum of Al_xO_y-200 exhibits NH stretching modes in the range of 3100–3500 cm^{-1} (Fig. 7c), originating from the amine groups of the urea ligands, and in accord with the behaviour of the nitrate-containing yttrium compound **2**. Additionally, these findings are in agreement with a DSC measurement of **3** displaying a sharp exothermic peak at 225 °C (Fig. 3d) as well as the thermogravimetric analysis of **3**, exhibiting a most prominent and significant mass loss at about 225 °C. As a consequence, capacitor devices processed at 200 °C are still exhibiting electrical short-circuiting throughout, indicating a significant presence of parasitic electrical pathways, while processing temperatures of 250–350 °C already lead to amorphous aluminium oxide, based capacitor devices with good dielectric properties.

Dielectric properties of solution derived yttrium oxide Y_xO_y and aluminium oxide Al_xO_y

The dielectric properties of the solution-processed metal oxide thin films processed at various temperatures were measured using a metal–insulator–metal (MIM) structure (Fig. 11).

ITO-coated glass was used for the fabrication of the capacitor devices, serving as a substrate as well as a gate electrode. Precursor solutions of **1**, **2** and **3** were spin-coated on the ITO coated glass and annealed at various temperatures (200–350 °C), leading to the formation of Y_xO_y and Al_xO_y thin films, respectively. As top electrode, gold (50 nm) was sputtered on the Y_xO_y films. In case of the Al_xO_y films a combination of 50 nm gold as well as a 10 nm interlayer of titanium was used as the top electrode.

The capacitance *vs.* frequency curves in the range of 10 Hz to 100 kHz are displayed in Fig. 12a–c. The use of precursor **1** thermally processed at 200–350 °C leads to Y_xO_y based capacitors, exhibiting areal capacities of 26, 41, 63 and 84 nF cm^{-2} at 10 kHz for Y_xO_y-(**1**)-200, Y_xO_y-(**1**)-250, Y_xO_y-(**1**)-300 and Y_xO_y-(**1**)-350, respectively. The use of precursor **2**, at an annealing temperature of 300 and 350 °C, results in a capacitance of 111 and 131 nF cm^{-2} at 10 kHz, respectively. Furthermore, it is noteworthy that the use of the yttrium precursors Y-DEM-NO$_2$ **1** and Y-UN **2** lead to capacitors with almost no frequency dispersion.

Finally, the Al_xO_y dielectric, obtained by thermal processing from precursor **3**, exhibit remarkable high capacities of 184, 216 and 259 nF cm^{-2} at 10 kHz for Al_xO_y-250, Al_xO_y-300 and Al_xO_y-350 respectively. The observed high capacities C can be attributed to a progressing conversion of **3** into the respective oxidic films of $\sim$ 60 nm thickness (Fig. 10) based on the relationship $C = (k E_0 A)/D$. Additionally, breakdown measurements have been carried out for all three dielectric layers (Fig. 12d–f). It is evident that higher annealing temperatures correspond to higher breakdown voltages as well as less current leakage. The samples Y_xO_y-(**1**)-250, Y_xO_y-(**1**)-300 and Y_xO_y-(**1**)-350, generated from **1** show no electrical breakdown throughout the whole measurement range of 40 V. Interestingly, the Y_xO_y-200 sample exhibit no clear cut electrical breakdown, but a drastic increasing leakage current with increasing voltage. This might be attributed to the fact that almost no yttrium oxide has formed at 200 °C, as shown in the XPS measurement (Fig. 8a). The leakage current at 1 MV cm^{-1}, using precursor **1** amounts to 5.5×10^{-8}, 6.9×10^{-10}, 2.1×10^{-10} and 3.8×10^{-10} A cm^{-2} for samples Y_xO_y-(**1**)-200, Y_xO_y-(**1**)-250, Y_xO_y-(**1**)-300 and Y_xO_y-(**1**)-350,

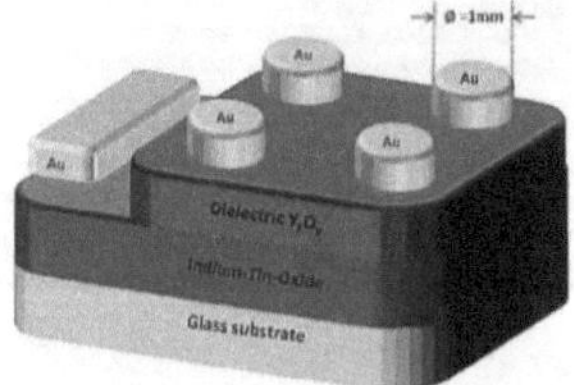

Fig. 11 Schematic illustration of the prepared capacitor. The grey layer represents the glass substrate, the purple layer represents a 140 nm ITO film serving as the bottom gate electrode and the blue layer illustrates the Y_xO_y dielectric film. The circular top electrodes are composed of 50 nm gold and on the left-hand side is a 100 nm gold sacrificial contact.

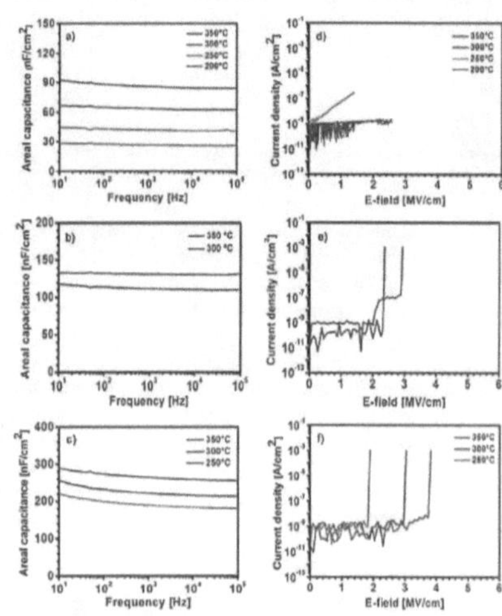

Fig. 12 Capacitance *vs.* frequency curves of solution processed Y_xO_y and Al_xO_y generated from (a) **1**, (b) **2** and (c) **3** annealed at different temperatures. Leakage current density *vs.* electric field behavior of Y_xO_y and Al_xO_y generated from (d) **1**, (e) **2** and (f) **3** annealed at different temperatures.

annealed at these different temperatures, respectively. Y_xO_y-(2)-300 and Y_xO_y-(2)-350 generated from precursor **2** possess an electrical breakdown of 2.4 and 2.9 MV cm^{-1} and a very low leakage current of 9.2×10^{-10} and 8.5×10^{-10} A cm^{-2} at 1 MV cm^{-1}. The electrical breakdown of the Al_xO_y dielectrics occur at 1.9, 3.0 and 3.8 MV cm^{-1} for Al_xO_y-250, Al_xO_y-300 and Al_xO_y-350, respectively. The current leakage at 1 MV cm^{-1} amounts 9.1×10^{-10}, 6.2×10^{-10} and 5.0×10^{-10} for Al_xO_y-250, Al_xO_y-300 and Al_xO_y-350, respectively. It is remarkable that besides Y_xO_y-(1)-200, all fabricated capacitors fulfil the criteria of Wager *et al.* for an ideal dielectric, exhibiting low leakage current of $J < 1 \times 10^{-8}$ at 1 MV cm^{-1}.[33]

The control of the atmospheric conditions under which the studies have been performed is crucial for the interpretation of the obtained dielectric properties. In ambient atmosphere, adsorbed water molecules reside on the thin film surface, contributing to the parasitic resistance and thus resulting in a drastically increased capacitance. At low annealing temperatures there are more detrimental hydroxide groups residing on the surface of the thin films, which enhance the adsorption of additional water molecules. Consequently, this effect becomes even higher; the lower the chosen annealing temperature under ambient conditions is. As a result, extraordinarily high capacities even for very low annealing temperatures are a clear indicator that such studies have been performed under ambient, atmospheric conditions. Therefore, a direct comparison of studies performed under non-inert and inert atmosphere seems unrealistic. The situation can be even more complicated if conditions of measurement are not explicitly comparable in this regard. An overview of the current state of the art studies is given in Table 1. It is evident that the yttrium oxide Y_xO_y dielectric derived from the molecular precursors **1** and **2** show very good electrical performances compared to other yttrium oxide-based dielectrics. Only work by Adamopolous *et al.* reports higher capacities (133 nF cm^{-2}) comparable values to those of Y_xO_y generated from precursor **2** (131 nF cm^{-2}), though using even a slightly higher annealing temperature of 400 °C. Concerning the

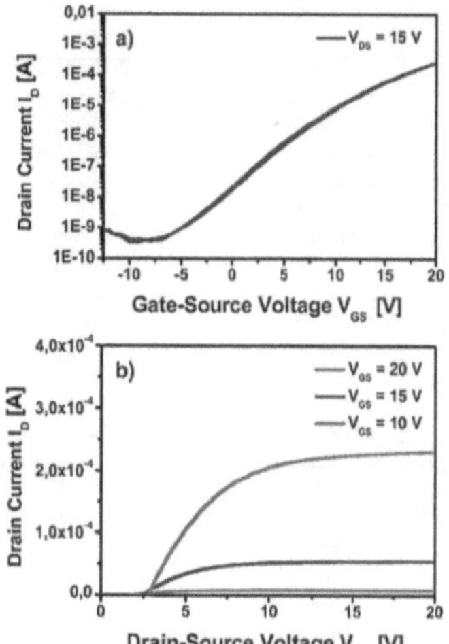

Fig. 13 Electrical characterization of TFT based on the Y_xO_y-(1)-350 dielectric and a solution processed indium zinc oxide (IZO) semiconductor processed at 350 °C. (a) Transfer characteristics of the device at $V_{DS} = 15$ V. (b) Output characteristics of the device measured at 10 V, 15 V and 20 V, respectively.

current leakage at 1 MV cm^{-1} the reported reference works display values which are up to three magnitudes higher compared to the current leakage generated from the herein reported yttrium oxide obtained from our precursor approach using molecules **1** and **2**. Indeed, the new capacitors reported herein are to the best of our

Table 1 Overview and comparison of dielectric properties of solution-processed Y_xO_y dielectric films

Ref.	Temp. (°C)	d (nm)	k	Areal cap. (nF cm^{-2}) (f)	Leakage (A cm^{-2})	E_B (MV cm^{-1})
This work 1	350	156	14.9	84 (10 kHz)	3.8×10^{-10} (1 MV cm^{-1})	>2
This work 1	300	163	11.5	63 (10 kHz)	2.1×10^{-10} (1 MV cm^{-1})	>2
This work 1	250	248	11.5	46 (10 kHz)	6.9×10^{-10} (1 MV cm^{-1})	>2
This work 1	200	286	8.5	21 (10 kHz)	5.5×10^{-8} (1 MV cm^{-1})	—
This work 2	350	82	12.1	131 (10 kHz)	8.5×10^{-10} (1 MV cm^{-1})	2.9
This work 2	300	87	10.9	111 (10 kHz)	9.2×10^{-10} (1 MV cm^{-1})	2.4
Song	500	188	15.9	74.7 (1 MHz)	8.63×10^{-7} (2 MV cm^{-1})	—
Song	400	188	15.6	73.4 (1 MHz)	7.21×10^{-8} (2 MV cm^{-1})	—
Song	300	188	15.2	71.7 (1 MHz)	5.24×10^{-8} (2 MV cm^{-1})	—
Tsay	550	220	10.5	42.2 (100 kHz)	1.7×10^{-6} (5 V)	—
Tsay	500	220	10.0	≈40 (100 kHz)	1.8×10^{-7} (5 V)	—
Tsay	450	225	9.5	37.4 (100 kHz)	$<1.0 \times 10^{-6}$ (5 V)	—
Adamopolous	400	—	≈16.2	≈133 (120 Hz)	—	—
[a]This work 3	350	59	17.2	259 (10 kHz)	5.0×10^{-10} (1 MV cm^{-1})	3.8
[a]This work 3	300	61	14.9	216 (10 kHz)	6.2×10^{-10} (1 MV cm^{-1})	3.0
[a]This work 3	250	74	15.4	184 (10 kHz)	9.1×10^{-10} (1 MV cm^{-1})	1.9

[a] Our solution processed Al_xO_y dielectric is given for comparison.

knowledge currently the only ones fulfilling the criterion of Wager *et al.* ($J < 1 \times 10^{-8}$ A cm^{-2} at 1 MV cm^{-1}) for minimum current leakage of technologically relevant dielectrics. Lastly, in order to evaluate the dielectric in an TFT geometry Y$_x$O$_y$ derived from precursor compound **1** was succesfully employed as a functional dielectric layer in a field effect transistor using solution-processed indium zinc oxide (IZO) as semiconducting layer. Within the device architecture, solution deposited amorphous indium zinc oxide (IZO) precursor was spin-coated and subsequently processed at 350 °C for complete formation of a IZO semiconductor. The IZO precursor solution was generated by using single source zinc and indium precursors as described by us.[22,44-46] A schematic representation of the TFT architecture is shown in Fig. S18 ESL† The electrical measurements are characterized by plotting the transfer and output characteristics of the measured TFT device and is displayed in Fig. 13. The crucial performance metrics including the charge-carrier mobility (μ_{sat}) in the saturation regime, the threshold voltage (V_{th}) and the ratio of the current measured in the on-state and the off-state ($I_{on/off}$) of the final device were extracted from the data obtained from the transfer and output characteristics. The measured TFT displays a good device performance with a μ_{sat} of 2.1 cm^2 V^{-1} s^{-1}, a V_{th} of 6.9 V and an $I_{on/off}$ of 7.6 × 10^5, with no significant hysteresis in the measured transfer characteristics. The device performance demonstrates the successful implementation of the fabricated Y$_x$O$_y$ dielectric when combined with an amorphous IZO semiconductor in a TFT device architecture thus serving as a proof of principle study.

Conclusions

In conclusion, we have demonstrated the synthesis and full structural characterization of the single-source molecular precursors bis(diethyl-2-nitromalonato) nitrato yttrium(III) **1**, dinitrato tetra(urea) yttrium(III)-nitrate **2** and hexakis(urea) aluminium(III)-nitrate **3**. The thermal decomposition of these precursors has been investigated in depth revealing formation of volatile by-products in the course of the thermal conversion with no significant formation of stable intermediate decomposition products. Additionally, in all cases the conversion into the respective metal oxide occurs at relatively low decomposition temperatures which exemplifies the versatility of the combustion route using high energy precursors. While precursor **1** is soluble in methoxyethanol, precursor **2** and **3** have good solubility in water (>20%), which provides an eco-friendly approach to the synthesis of the high-*k* Y$_x$O$_y$ and Al$_x$O$_y$ dielectric materials. Al$_x$O$_y$ thin films are amorphous for all annealing temperatures studied (200–350 °C), whereby Y$_x$O$_y$ samples show crystallinity starting from 350 °C for **1** and 250 °C for **2**, respectively. The Al$_x$O$_y$ based capacitors fabricated exhibit very high areal capacities even at a moderate annealing temperature of 250 °C (184 nF cm^{-2} at 10 kHz). Furthermore, all Al$_x$O$_y$ based capacitors, annealed in the temperature range 250–350 °C, exhibit very low current leakage of $J < 10^{-9}$ A cm^{-2} at 1 MV cm^{-1}. Additionally, the breakdown voltage was determined to be 1.9, 3.0 and 3.8 MV cm^{-1} for Al$_x$O$_y$-250, Al$_x$O$_y$-300 and Al$_x$O$_y$-350, respectively.

The Y$_x$O$_y$ based capacitors, generated from precursor **2** and annealed at 300 and 350 °C exhibit a high capacitance of 111 and 131 nF cm^{-2} at 10 kHz. The current leakage at 1 MV cm^{-1} amounts less than 10^{-9} A cm^{-2} at 1 MV cm^{-1} for both samples and the electrical breakdown occurs at 2.4 and 2.9 MV cm^{-1}, respectively. The Y$_x$O$_y$ based capacitors, generated from precursor **1**, exhibit satisfactory areal capacities even at moderate temperatures (84 nF cm^{-2} at 350 °C). Furthermore, the Y$_x$O$_y$ based capacitors exhibit almost no frequency dispersion over the whole range (10 Hz–100 kHz). Except for a calcination temperature of 200 °C, all capacitors exhibit very low current leakage of $J < 10^{-9}$ A cm^{-2} at 1 MV cm^{-1} and show no electrical breakdown up to 40 V.

Additionally, the implementation of the fabricated yttrium oxide dielectric with an IZO semiconductor processed at 350 °C demonstrates an effective TFT performance characteristics, exhibiting a μ_{sat} of 2.1 cm^2 V^{-1} s^{-1}, a V_{th} of 6.9 V and $I_{on/off}$ of 7.6 × 10^5.

The ability of such a unique class of high combustible precursor molecules based on the urea/nitrate ligand environment presents itself as a potential alternative in the low-temperature fabrication of high-*k* dielectric materials. It remains interesting to see if this strategy can be extended towards formation of other metal oxides from a similar solution process. This could warrant interesting functional properties towards novel device fabrication in the future.

Conflicts of interest

The authors declare no conflict of interests.

Acknowledgements

TEM investigations were performed at ERC Jülich under contract with ERC-TUD1. We acknowledge Dr J. Engstler and S. Heinschke for performing TEM, XRD and ellipsometric studies. The acquisition of the K-Alpha(+) instrument at KIT was supported by the German Federal Ministry of Economics and Technology.

Notes and references

1 A. Facchetti and T. J. Marks, *Transparent Electronics: From Synthesis to Applications*, 2010.

2 S. Park, C.-H. Kim, W.-J. Lee, S. Sung and M.-H. Yoon, *Mater. Sci. Eng., R*, 2017, **114**, 1–22.

3 J. Robertson and R. M. Wallace, *Mater. Sci. Eng., R*, 2015, **88**, 1–41.

4 W. Xu, H. Wang, L. Ye and J. Xu, *J. Mater. Chem. C*, 2014, **2**, 5389–5396.

5 R. C. Frunză, B. Kmet, M. Jankovec, M. Topič and B. Malič, *Mater. Res. Bull.*, 2014, **50**, 323–328.

6 Y. B. Yoo, J. H. Park, K. H. Lee, H. W. Lee, K. M. Song, S. J. Lee and H. K. Baik, *J. Mater. Chem. C*, 2013, **1**, 1651–1658.

7 C.-G. Lee and A. Dodabalapur, *J. Electron. Mater.*, 2012, **41**, 895–898.

8 A. Liu, G. X. Liu, H. H. Zhu, F. Xu, E. Fortunato, R. Martins and F. K. Shan, *ACS Appl. Mater. Interfaces*, 2014, **6**, 17364–17369.

9 L. Zhu, G. He, J. Lv, E. Fortunato and R. Martins, *RSC Adv.*, 2018, **8**, 16788–16799.

10 C. Avis and J. Jang, *J. Mater. Chem.*, 2011, **21**, 10649–10652.

11 Y. Xu, X. Li, L. Zhu and J. Zhang, *Mater. Sci. Semicond. Process.*, 2016, **46**, 23–28.

12 W. Yang, K. Song, Y. Jung, S. Jeong and J. Moon, *J. Mater. Chem. C*, 2013, **1**, 4275–4282.

13 P. K. Nayak, M. N. Hedhili, D. Cha and H. N. Alshareef, *Appl. Phys. Lett.*, 2013, **103**, 033518.

14 K. Song, W. Yang, Y. Jung, S. Jeong and J. Moon, *J. Mater. Chem.*, 2012, **22**, 21265–21271.

15 A. Liu, G. Liu, H. Zhu, Y. Meng, H. Song, B. Shin, E. Fortunato, R. Martins and F. Shan, *Curr. Appl. Phys.*, 2015, **15**, S75–S81.

16 C.-Y. Tsay, C.-H. Cheng and Y.-W. Wang, *Ceram. Int.*, 2012, **38**, 1677–1682.

17 F. Xu, A. Liu, G. Liu, B. Shin and F. Shan, *Ceram. Int.*, 2015, **41**, S337–S343.

18 G. Adamopoulos, S. Thomas, D. D. C. Bradley, M. A. McLachlan and T. D. Anthopoulos, *Appl. Phys. Lett.*, 2011, **98**, 123503.

19 W. Xu, H. Li, J.-B. Xu and L. Wang, *ACS Appl. Mater. Interfaces*, 2018, **10**, 25878–25901.

20 A. Liu, H. Zhu, H. Sun, Y. Xu and Y.-Y. Noh, *Adv. Mater.*, 2018, **30**, 1706364.

21 S. W. Smith, W. Wang, D. A. Keszler and J. F. Conley Jr, *J. Vac. Sci. Technol., A*, 2014, **32**, 041501.

22 N. Koslowski, S. Sanctis, R. C. Hoffmann, M. Bruns and J. J. Schneider, *J. Mater. Chem. C*, 2019, **7**, 1048–1056.

23 G. Teowee, K. C. McCarthy, F. S. McCarthy, T. J. Bukowski, D. G. Davis and D. R. Uhlmann, *J. Sol-Gel Sci. Technol.*, 1998, **13**, 895–898.

24 H. Z. Zhang, L. Y. Liang, A. H. Chen, Z. M. Liu, Z. Yu, H. T. Cao and Q. Wan, *Appl. Phys. Lett.*, 2010, **97**, 122108.

25 W. Xu, M. Long, T. Zhang, L. Liang, H. Cao, D. Zhu and J.-B. Xu, *Ceram. Int.*, 2017, **43**, 6130–6137.

26 A. Liu, G. Liu, H. Zhu, B. Shin, E. Fortunato, R. Martins and F. Shan, *RSC Adv.*, 2015, **5**, 86606–86613.

27 W. Xu, H. Wang, F. Xie, J. Chen, H. Cao and J.-B. Xu, *ACS Appl. Mater. Interfaces*, 2015, **7**, 5803–5810.

28 H. Tan, G. Liu, A. Liu, B. Shin and F. Shan, *Ceram. Int.*, 2015, **41**, S349–S355.

29 D.-H. Lee, Y.-J. Chang, W. Stickle and C.-H. Chang, *Solid-State Lett.*, 2007, **10**, K51–K54.

30 Y. Zhao, G. Dong, L. Duan, J. Qiao, D. Zhang, L. Wang and Y. Qiu, *RSC Adv.*, 2012, **2**, 5307–5313.

31 B. Sykora, D. Wang and H. v. Seggern, *Appl. Phys. Lett.*, 2016, **109**, 033501.

32 R. C. Hoffmann and J. J. Schneider, *Eur. J. Inorg. Chem.*, 2014, **2014**, 2241–2247.

33 Y. Chen, B. Wang, W. Huang, X. Zhang, G. Wang, M. J. Leonardi, Y. Huang, Z. Lu, T. J. Marks and A. Facchetti, *Chem. Mater.*, 2018, **30**, 3323–3329.

34 S. Sanctis, R. C. Hoffmann, M. Bruns and J. J. Schneider, *Adv. Mater. Interfaces*, 2018, **5**, 1800324.

35 R. C. Hoffmann, M. Kaloumenos, S. Heinschke, E. Erdem, P. Jakes, R.-A. Eichel and J. J. Schneider, *J. Mater. Chem. C*, 2013, **1**, 2577–2584.

36 S. Sanctis, N. Koslowski, R. Hoffmann, C. Guhl, E. Erdem, S. Weber and J. J. Schneider, *ACS Appl. Mater. Interfaces*, 2017, **9**, 21328–21337.

37 Y. Qiu and L. Gao, *J. Am. Ceram. Soc.*, 2004, **87**, 352–357.

38 M. S. Lupin and G. E. Peters, *Thermochim. Acta*, 1984, **73**, 79–87.

39 S. Sanctis, R. C. Hoffmann, N. Koslowski, S. Foro, M. Bruns and J. J. Schneider, *Chem. - Asian J.*, 2018, **13**, 3912–3919.

40 A. S. Antsyshkina, G. G. Sadikov, M. N. Rodnikova and S. E. Tikhonov, *Crystallogr. Rep.*, 2003, **48**, 610–612.

41 H. A. Al-Abadleh and V. H. Grassian, *Langmuir*, 2003, **19**, 341–347.

42 J. Gangwar, B. K. Gupta, P. Kumar, S. K. Tripathi and A. K. Srivastava, *Dalton Trans.*, 2014, **43**, 17034–17043.

43 D. Barreca, G. A. Battiston, D. Berto, R. Gerbasi and E. Tondello, *Surf. Sci. Spectra*, 2001, **8**, 234–239.

44 C. Durand, C. Dubourdieu, C. Vallée, V. Loup, M. Bonvalot, O. Joubert, H. Roussel and O. Renault, *J. Appl. Phys.*, 2004, **96**, 1719–1729.

45 T. Gougousi and Z. Chen, *Thin Solid Films*, 2008, **516**, 6197–6204.

46 E. J. Rubio, V. V. Atuchin, V. N. Kruchinin, L. D. Pokrovsky, I. P. Prosvirin and C. V. Ramana, *J. Phys. Chem. C*, 2014, **118**, 13644–13651.

47 J. Stoch and J. Gablankowska-Kukucz, *Surf. Interface Anal.*, 1991, **17**, 165–167.

48 P. Post, L. Wurlitzer, W. Maus-Friedrichs and A. P. Weber, *Nanomaterials*, 2018, **8**, 530.

49 M. Y. Smirnov, A. V. Kalinkin and V. I. Bukhtiyarov, *J. Struct. Chem.*, 2007, **48**, 1053–1060.

50 B. C. Beard, *Appl. Surf. Sci.*, 1990, **45**, 221–227.

51 V. Trouillet, H. Tröße, M. Bruns, E. Nold and R. White , *J. Vac. Sci. Technol. A*, 2007, **25**, 927–931.

52 B. R. Strohmeier, *Surf. Sci. Spectra*, 1994, **3**, 135–140.

53 C. Gao, X.-Y. Yu, R.-X. Xu, J.-H. Liu and X.-J. Huang, *ACS Appl. Mater. Interfaces*, 2012, **4**, 4672–4682.

54 M. R. Alexander, G. E. Thompson, X. Zhou, G. Beamson and N. Fairley, *Surf. Interface Anal.*, 2002, **34**, 485–489.

55 J. F. Wager, D. A. Keszler and R. E. Presley, *Transparent Electronics*, Springer, New York, 2008.

6.3 Solution-processed amorphous yttrium aluminium oxide YAl$_x$O$_y$ and aluminum oxide Al$_x$O$_y$ and their functional dielectric properties and performance in thin-film transistors

This publication is focusing on the formation of the ternary oxide dielectric YAl$_x$O$_y$, consisting of aluminum oxide and yttrium oxide. We decided to study ternary oxides due to the fact that the majority of binary oxide dielectrics suffer from an inverse trend between the amount of the dielectric constant (k) and the band gap energy (E$_G$), resulting in compromised overall device performances. Our strategy to overcome this issue is the generation of hybrid multinary oxide dielectrics, e.g. ternary oxide dielectrics, wherein one material possess a high dielectric constant and the other one provides a wide band gap, resulting in a dielectric material with a sufficiently high dielectric constant as well as a decreased leakage current density and thus exhibiting superior electrical properties in comparison to the individual binary oxides.

We utilized the previously established single-source molecular precursors tris[(diethyl-2-nitromalonato)]aluminum(III), Al-DEM-NO$_2$ and bis(diethyl-2-nitromalonato) nitrato yttrium(III), Y-DEM-NO$_2$ for the formation of the ternary metal oxide dielectric yttrium aluminium oxide YAl$_x$O$_y$. Thereby, thermal (350 °C) as well as DUV-mediated low-temperature (150 °C) solution-processing of various YAl$_x$O$_y$ compositions was performed due to the strong absorption of these precursors at wavelengths of λ < 250 nm. The YAl$_x$O$_y$-350 samples of all investigated compositions are amorphous throughout, even on the nanometer scale, exhibiting a very smooth surface roughness of R$_{RMS}$ < 0.2 nm. The incorporation of yttrium leads to a non-linear behavior of the dielectric film thickness, which is attributable to the inclusion of yttrium into the Al$_x$O$_y$ host lattice, reaching the saturation at an yttrium incorporation of 30 mol-%. The XPS surface analysis reveals that the incorporation of yttrium into the amorphous Al$_x$O$_y$ lattice results in a decrease of the carbonate species within the material (10-YAl$_x$O$_y$, 30-YAl$_x$O$_y$), corresponding to a decreased leakage current density of the final capacitor device. Thereby, all YAl$_x$O$_y$-350 based capacitors possess a very low leakage current density of J < 2 x 10^{-9} A cm^{-2} at 1 MV cm^{-1}. Additionally, for all YAl$_x$O$_y$ compositions no electrical breakdown occurs up to E$_B$ > 3.3 MV cm^{-1}. Besides, all YAl$_x$O$_y$-350 based capacitors exhibit increased dielectric constants of k = 12.0, 12.5 and 13.3 at 10 kHz for 10-YAl$_x$O$_y$, 30-YAl$_x$O$_y$ and 50-YAl$_x$O$_y$, respectively. In addition, almost no frequency dispersion (10 Hz–100 kHz) is observable from an yttrium incorporation $\geq$ 30 mol-%. Finally, TFT devices based on the solution-processed 30-YAl$_x$O$_y$-350 dielectric and combined with an active IZO semiconductor exhibit reasonable performance characteristics with a μ_{sat} value of 2.6 cm^2 V^{-1} s^{-1}, V$_{on}$ of -1.1 V, V$_{th}$ of 12.4 V and I$_{on/off}$ of 1.8 x 10^7.

Regarding the DUV-processed dielectrics, the yttrium precursor Y-DEM-NO$_2$ does not undergo the desirable photolytic decomposition at a wavelength of λ = 160 nm, resulting in poor YAl$_x$O$_y$ based dielectrics. Nevertheless, the DUV-mediated Al$_x$O$_y$ dielectric itself displays excellent dielectric properties (k = 9.0, C$_i$ = 153 nF cm^{-2}, J = 1.7 x 10^{-9} A cm^{-2} at 1 MV cm^{-1} and E$_B$ = 4.1 MV cm^{-1}).

Journal of
Materials Chemistry C

ROYAL SOCIETY
OF CHEMISTRY

PAPER

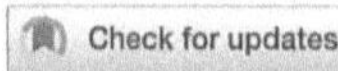 Check for updates

Cite this: *J. Mater. Chem. C*, 2020, 8, 8521

Solution-processed amorphous yttrium aluminium oxide YAl$_x$O$_y$ and aluminum oxide Al$_x$O$_y$, and their functional dielectric properties and performance in thin-film transistors†

Nico Koslowski, [a] Vanessa Trouillet[b] and Jörg J. Schneider *[a]

Fabrication of dielectric yttrium aluminium oxide YAl$_x$O$_y$ thin films utilizing nitro functionalized malonato complexes of aluminium Al-DEM-NO$_2$ **1** and yttrium Y-DEM-NO$_2$ **2** is described. Their controlled thermal conversion under moderate conditions into amorphous ternary yttrium aluminium oxide dielectric thin films with a very low surface roughness ($R_{RMS} < 0.2$ nm) was accomplished. Capacitor devices composed of YAl$_x$O$_y$ exhibit excellent dielectric properties at a processing temperature of 350 °C. It was possible to systematically modulate their significant performance parameters such as dielectric constant, leakage current density and breakdown voltage according to the amount of yttrium rare earth metal incorporation which showed an optimum at 30 mol% yttrium incorporation. The yttrium inclusion into the amorphous Al$_x$O$_y$ lattice reduces the formation of carbonate species in comparison to yttrium free pristine Al$_x$O$_y$ dielectric thin films. Finally, a solution-processed thin-film transistor (TFT) with YAl$_x$O$_y$ (30 mol% – Y content) as the dielectric and indium zinc oxide (IZO) as the semiconductor exhibits good TFT characteristics with a saturation mobility (μ_{sat}) of 2.6 cm^2 V^{-1} s^{-1}, an on-voltage (V_{on}) of -1.1 V, a threshold voltage (V_{th}) of 12.4 V and an on/off current ratio ($I_{on/off}$) of 1.8×10^7. As an alternative to a necessary thermal annealing step deep UV irradiation at 160 nm allows the implementation of low-temperature (150 °C) solution processing of a yttrium free pure Al$_x$O$_y$ dielectric, which exhibits excellent dielectric performance parameters. It displays an areal capacity of 153 nF cm^{-2}, a leakage current density of 1.7×10^{-9} A cm^{-2} at 1 MV cm^{-1} and a breakdown voltage of 4.1 MV cm^{-1}.

Received 16th April 2020,
Accepted 19th May 2020

DOI: 10.1039/d0tc01876g

rsc.li/materials-c

Introduction

High-κ dielectrics based on metal oxides are potential candidates as the gate dielectric in oxide based thin-film transistor (TFT) devices, and are close to replace SiO$_2$.[1,2] For the development of high performance TFT devices, the choice of the gate dielectric is of great importance. The switching performance of transistor devices is highly dependent on the amount of charges accumulated in the semiconductor material, which depends on the degree of disorder within the semiconductor as well as the electrical polarization capacity of the dielectric.[3,4] Thereby, the use of high-κ dielectrics can result in TFT devices with higher mobilities and lower threshold voltages in comparison to

TFTs utilizing the established SiO$_2$ dielectrics.[5,6] Amorphous thin films obtained *via* spin-coating or printing techniques with smooth surfaces are highly desirable regarding TFT applications due to an improved dielectric/semiconductor interface, resulting in lower leakage currents and better reliability.[7,8] Currently, various binary metal oxides such as TiO$_2$,[9] ZrO$_2$,[9–12] HfO$_2$,[13] Ta$_2$O$_5$,[14] Y$_2$O$_3$[9,15–20] and Al$_2$O$_3$[9,20–25] have already shown very good results with respect to their dielectric properties in such devices, but all of these binary oxide dielectrics suffer from a significant drawback. An ideal dielectric material should provide a sufficiently high dielectric constant ($\kappa > 10$), therefore displaying a high areal capacitance, while exhibiting a low leakage current density ($J < 1 \times 10^{-8}$ A cm^{-2}).[26] Unfortunately, binary oxide dielectrics with higher dielectric constants usually possess narrow band gaps and thus suffer from high leakage currents, which may deteriorate the overall device performance. One possible strategy to overcome this issue is the generation of hybrid multinary oxide dielectrics, *e.g.* ternary oxide dielectrics, wherein one dielectric material may possess a high dielectric constant and the other one provides a wide band gap, resulting in a decreased leakage current and thus exhibiting superior

a Fachbereich Chemie, Eduard-Zintl-Institut für Anorganische und Physikalische Chemie, Technische Universität Darmstadt, Alarich-Weiss-Str. 12, 64287 Darmstadt, Germany. E-mail: joerg.schneider@tu-darmstadt.de
b Institute for Applied Materials (IAM-ESS) and Karlsruhe Nano Micro Facility (KNMF), Karlsruhe Institute of Technology (KIT), Hermann-von-Helmholtz-Platz 1, 76344 Eggenstein-Leopoldshafen, Germany

† Electronic supplementary information (ESI) available. See DOI: 10.1039/d0tc01876g

electrical properties in comparison to the individual binary oxides. According to this, a number of ternary oxide dielectrics such as La_xAl_yO,[27] Zr_xAl_yO,[28] Hf_xAl_yO,[29] Hf_xSi_yO,[30] Al_xP_yO[31] and Al_xNa_yO[32] have been subsequently investigated. Another promising candidate as a gate dielectric in the realm of TFT devices is yttrium aluminium oxide YAl_xO_y. Therein Y_2O_3 represents the high-κ material ($\kappa \approx 15$)[33] component and Al_2O_3 the material with a wide band gap ($E_g \approx 8.8$ eV).[34] Consequently, research in this field has especially gained impetus in the last two years. In 2018 Jeong et al. reported on solution-processed yttrium aluminium oxide thin films with excellent dielectric properties, fabricated at an annealing temperature of 400 °C.[35] In this work they systematically studied three different compositions of yttrium aluminium oxide dielectrics and compared them to the respective binary oxides of aluminium and yttrium. All the YAl_xO_y compositions obtained displayed higher dielectric constants compared to the pure binary Al_xO_y dielectric. In addition, all YAl_xO_y dielectrics exhibited higher breakdown voltages and interestingly also lower leakage current densities in comparison to both the pristine Y_xO_y and Al_xO_y dielectrics. Jeong et al. attribute the decrease of the leakage current density in such ternary hybrid materials to the decrease of hydroxyl species within the YAl_xO_y layers, which is related to the amount of yttrium incorporation into the Al_xO_y lattice. Subramanian et al. successfully reduced the post deposition annealing (PDA) temperature (250 °C) for the fabrication of printed Al_xO_y, with decent dielectric properties, by introducing a pre-treatment with UV radiation ($\lambda > 350$ nm). Furthermore, it was possible to fabricate well performing TFT devices using a printed In_xO_y based semiconductor.[36] In 2019, Bolat and Romanyuk et al. expanded this approach to the ternary metal oxide dielectric yttrium aluminium oxide. Processing of the $YAlO_x$ dielectric was achieved by what these authors called deep-UV (DUV) annealing ($\lambda < 260$ nm) at temperatures as low as 150 °C.[37] Besides the incremental increase of the dielectric constant, the incorporation of yttrium resulted in a decreased leakage current density as already observed by Jeong et al. Furthermore, Bolat and Romanyuk et al. successfully demonstrated the implementation of the UV-mediated $YAlO_x$ dielectric within a TFT device by means of inkjet printing on flexible substrates. Therefore the UV-mediated solution processing of metal oxides presents a promising alternative to the widely used thermal processing when using suitable precursors which allow decomposition into active oxide layers by lowering the required fabrication temperature drastically.[38] As a consequence, the use of UV irradiation represents a promising novel approach towards the application of solution-processed TFT devices fabricated on flexible plastic substrates.

Currently, most of the conventional solution processing approaches utilize metal salts (nitrates, chlorides and acetates) as precursors, which usually exhibit some significant drawbacks such as the requirement of higher decomposition temperatures in order to completely decompose the precursor in comparison to molecular single-source precursors used herein.[39] Another issue is the necessary requirement of additives or stabilizers for such precursor solutions, which often renders the use of traditional inorganic sources for the metal oxides complicated, resulting in uncertain film formation processes with hardly controllable thin film morphology and composition, thus leading to unreliable electrical properties of the final device.[40]

In this work we have chosen our previously developed molecular single-source precursors tris[[(diethyl-2-nitromalonato)]-aluminium(III), Al-DEM-NO₂ **1**, and bis[[(diethyl-2-nitromalonato)]-nitrato yttrium(III), Y-DEM-NO₂ **2**, for a systematic study of the formation of amorphous thin films of the ternary oxide YAl_xO_y (yttrium aluminium oxide).[20,21] The coordination compounds **1** and **2** have already proven their feasibility as precursors for the thin film formation of binary oxides of aluminium and yttrium dielectrics with excellent dielectric properties utilizing a decent processing temperature of 350 °C. Adjusting the film compositions in a systematic manner and processing the compositions at 350 °C allows us to control significant dielectric characteristics such as areal capacitance and leakage current density, thus representing an effective approach to optimize these gate dielectrics in TFT devices.

Furthermore, both precursors display an intense absorption at wavelengths of $\lambda < 250$ nm, which makes them suitable as precursors for non-thermal UV-mediated solution processing of metal oxide dielectrics, also studied in the current work.

Experimental section

Synthesis of the metal oxides YAl_xO_y and Al_xO_y

The synthesis of the aluminium and yttrium precursors **1** and **2** with diethyl-2-nitromalonate ligands was performed as reported.[20,21] The respective metal oxide thin films were solution-processed using the spin-coating technique. Solutions of the respective precursors **1** and **2** in 2-methoxyethanol (20 wt%) were spin-coated on the substrate (3000 rpm, 20 s) and subsequently annealed at 350 °C or alternatively the deposited layers were dried at 150 °C and afterwards treated with DUV radiation by using a Hamamatsu L11798 H2D2 light source unit with an integrated air cooling system ($T < 50$ °C) and a MgF_2 window, emitting a spectral distribution of 115–400 nm, with the prominent main emission peak at 160 nm.

MIM capacitor fabrication

Solutions of the respective precursors **1** and **2** were prepared by dissolving 20 wt% of each in 2-methoxyethanol followed by subsequent sonication for 20 minutes and filtering the solution through a 0.2 mm polytetrafluoroethylene (PTFE) syringe filter. Afterwards, the solutions were mixed (using various Al/Y ratios) and stirred vigorously for two hours at 50 °C. ITO-coated glass substrates (140 nm, OLED grade) were used for the fabrication of the capacitors. First, the substrates were cleaned in deionized water, acetone and isopropanol for 15 minutes each using an ultrasonic bath. In order to enable electrical contact with the ITO layer, a 100 nm rectangular gold layer was sputter-deposited with the help of a shadow mask. The shadow mask exposes circular areas with diameters of 500, 750 and 1000 μm. Prior to spin-coating, the substrates were exposed to an air plasma for two minutes to enhance the hydrophilicity of the substrate.

Subsequently, the clear precursor solutions were spin-coated on the substrate (20 s; 3000 rpm) and annealed at 350 °C. This procedure was repeated three times and the thin film was finally annealed for two hours, resulting in amorphous and smooth films with thicknesses of 107–118 nm. The thickness of the layers was determined by spectroscopic ellipsometry. In all cases a 60 nm circular gold layer served as the top electrode. In order to investigate the contribution of yttrium to the final YAl_xO_y dielectric, three different compositions were used and abbreviated with respect to the yttrium concentration. For example, the dielectric with 50 mol% yttrium, processed at 350 °C, is abbreviated as 50-YAl_xO_y-350. Regarding the DUV-mediated samples, the clear precursor solutions were spin-coated on the substrate (20 s; 3000 rpm), followed by drying of the deposited layers at 150 °C. Afterwards the dried films were treated with DUV radiation with a maximum intensity at a wavelength of $\lambda = 160$ nm. The integrated air cooling system ensures a photolytic decomposition at $T < 50$ °C. The deposition procedure can be repeated until the desired film thickness is achieved. Finally, the thin films were treated with DUV radiation for one hour. The pristine DUV-processed Al_xO_y dielectric is abbreviated as Al_xO_y-DUV in the following. For the determination of the average dielectric characteristics 3 samples of each composition have been fabricated and 15 devices were evaluated.

Thin-film transistor (TFT) fabrication

In order to fabricate the TFT device, the indium zinc oxide (IZO) semiconductor was fabricated by using the established oximato precursor route using compounds of indium and zinc reported previously by our group.[20,21,41–46] Solutions of the respective indium and zinc precursors were prepared by dissolving 1 wt% of each in 2-methoxyethanol (In : Zn ratio, 6 : 4), followed by spin-coating of these solutions onto the 30-YAl_xO_y dielectric at 2500 rpm for 20 seconds and subsequent annealing at 350 °C. Performing three iterations of the coating procedure results in a film thickness of ~12 nm. Finally, gold source-drain electrodes (W/L = 2 mm/100 μm) were sputter-deposited onto the IZO semiconductor using a shadow mask (2 rectangular areas of 2 mm × 1 mm, distance of separation = 100 μm).

Material characterization

UV-Vis measurements of the precursor solutions were performed with a Thermo Scientific Evolution 600 using quartz cuvettes. X-ray diffraction (XRD) measurements were carried out on a MiniFlex 600 (Rigaku) using Cu Kα radiation (600 W) in Bragg–Brentano geometry. Transmission electron microscopy (TEM) was carried out with an operating voltage of 200 kV using a Tecnai F20 (FEI) system. The samples were prepared on lacey carbon-coated copper grids. Spectroscopic ellipsometry was performed using a Woollam M-2000 V spectrometer (spectral range 370–1690 cm^{-1}) using the CompleteEASE software (version 6.29). Atomic force microscopy (AFM) was performed using an MFP-3D (Asylum Research) equipped with silicon cantilevers. X-ray photoelectron spectroscopy (XPS) measurements were carried out on a K-Alpha(+) XPS system (Thermo Fisher Scientific, East Grinstead, UK). A monochromated Al Kα X-ray source with a spot size of 400 μm was used. The K-Alpha+ charge compensation system was used during the analysis, using electrons of 8 eV energy and low-energy argon ions to prevent any localized charge build-up. The spectra were fitted with one or more Voigt profiles (BE uncertainty: ±0.2 eV) and Scofield sensitivity factors were applied for quantification.[47] All spectra are referenced to the C 1s peak of hydrocarbons at 285.0 eV by means of the well-known photoelectron peaks of metallic Cu, Ag and Au, respectively.

Electrical characterization

Impedance measurements were performed using a ModuLab MTS System (Solartron Analytical Ltd) equipped with a probe station (Cascade Microtech, Inc) in a controlled argon atmosphere (O_2, H_2O < 0.5 ppm) in a glovebox. Impedance measurements were carried out in the frequency range of 10 Hz to 100 kHz with an amplitude of 500 mA. I–V measurements were performed on a B1500A semiconductor device analyzer (Agilent). TFT transfer and output characteristics were determined in a glove box in a controlled argon atmosphere (O_2, H_2O < 0.5 ppm) and in the dark using an HP 4155A semiconductor parameter analyzer (Agilent).

Results and discussion

Fig. 1 shows the UV-Vis spectra of the molecular single-source precursors Al-DEM-NO_2 1 and Y-DEM-NO_2 2. It is evident that these precursors possess a local absorbance maximum at 313 and 303 nm, respectively, which is in accordance with their yellowish colour. Furthermore, both precursors exhibit an additional strong absorbance at a lower wavelength of $\lambda < 250$ nm which renders them suitable for a DUV-mediated decomposition process. The X-ray analysis of the solution-processed decomposition products derived using various ratios of precursors 1 and 2 processed thermally at 350 °C (Fig. 2a) reveals completely amorphous YAl_xO_y products even on the nanoscopic scale (see Fig. 2b for the TEM image of the 50-YAl_xO_y sample). An amorphous state is highly desirable for a dielectric material

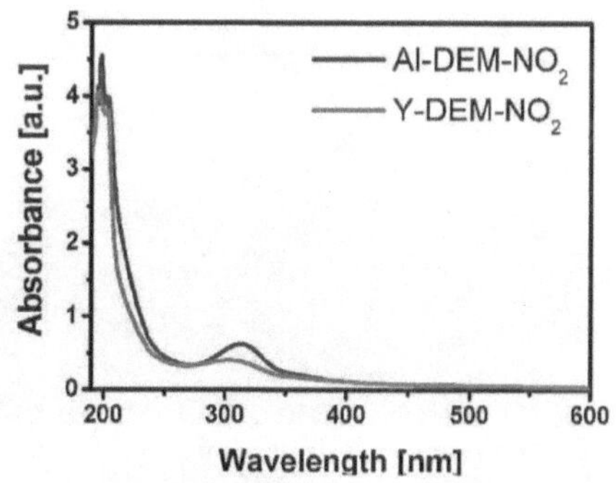

Fig. 1 UV-Vis spectra of the molecular single-source precursors Al-DEM-NO₂ 1 and Y-DEM-NO₂ 2.

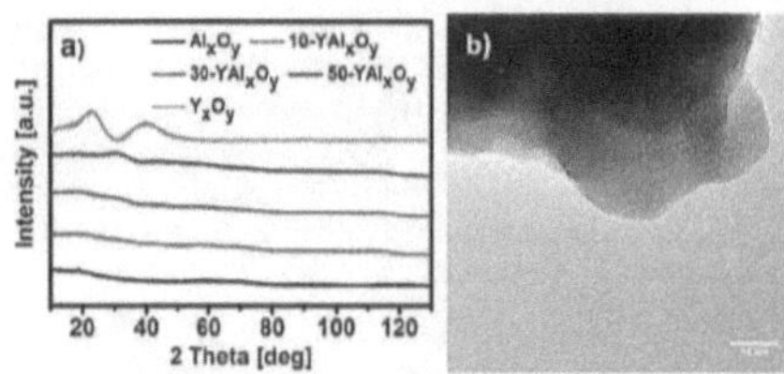

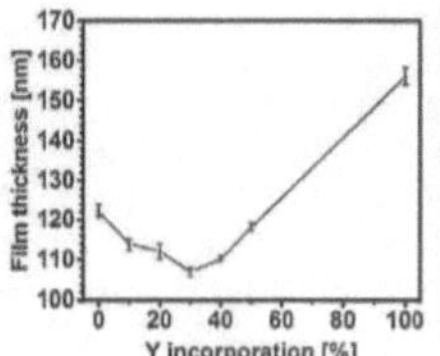

Fig. 2 (a) XRD patterns of different thermally processed compositions of YAl$_x$O$_y$ with varying yttrium/aluminium ratios together with Al$_x$O$_y$-350[21] and Y$_x$O$_y$-350[20] single phase dielectrics all derived from precursors **1** and **2** and annealed at 350 °C. (b) The TEM image of 50-YAl$_x$O$_y$ prepared at 350 °C.

Fig. 4 Film thicknesses of solution-processed dielectrics based on the yttrium incorporation. The films were annealed at 350 °C and the film thicknesses were determined by spectroscopic ellipsometry. The measurements were performed in an ambient atmosphere at 20 °C. The MSE values of the samples ranged from 6.4 to 9.0.

Sample	Thickness [nm]
Al$_x$O$_y$-350	122 ± 1.9
10-YAl$_x$O$_y$-350	114 ± 1.6
20-YAl$_x$O$_y$-350	112 ± 2.0
30-YAl$_x$O$_y$-350	107 ± 1.2
40-YAl$_x$O$_y$-350	110 ± 0.9
50-YAl$_x$O$_y$-350	118 ± 1.1
Y$_x$O$_y$-350	156 ± 2.2

because grain boundaries at crystalline interfaces can act as preferential pathways for impurity diffusion, leading to increased leakage current densities, thus resulting in an inferior performance of the desired dielectric material. Regarding the DUV-processed Al$_x$O$_y$ and Y$_x$O$_y$ samples the X-ray analysis revealed amorphous products (Fig. 3a). Again, their amorphous nature on the nanometer scale is confirmed by means of TEM (Fig. 3b and c).

In order to determine the film thickness of the various YAl$_x$O$_y$ compositions spectroscopic ellipsometry was performed. Interestingly, a non-linear behaviour of the film thickness with the increasing yttrium incorporation was observed. A systematic study of different film thicknesses (Fig. 4) was performed. The reason for the observed non-linear behaviour is the inclusion of yttrium into the amorphous Al$_x$O$_y$ lattice, which leads to an

increased density of the material and therefore thinner films in comparison to the pristine Al$_x$O$_y$-350[21] thin films. The addition of ~30 mol% yttrium (30-YAl$_x$O$_y$) results in the saturation of the yttrium inclusion into the Al$_x$O$_y$ lattice and therefore the sample displays the lowest film thickness (107 nm) of all the investigated compositions. Further addition of yttrium (40-YAl$_x$O$_y$ and 50-YAl$_x$O$_y$) beyond the saturation limit leads to an increase of the film thickness (110 and 118 nm) due to the growth of a separate Y$_x$O$_y$ phase.

Atomic force microscopy (AFM) was used to investigate the texture and morphology of the obtained dielectric thin films (Fig. 5). The YAl$_x$O$_y$ samples processed at 350 °C possessed a very smooth surface roughness, that is, R_{RMS} = 0.15, 0.11 and 0.12 nm for 10-YAl$_x$O$_y$, 30-YAl$_x$O$_y$ and 50-YAl$_x$O$_y$, respectively (Fig. 5a–c). Thereby, all YAl$_x$O$_y$ samples displayed a slightly smoother surface roughness compared to the single metal oxide dielectrics, exhibiting R_{RMS} = 0.35 and 0.31 nm for Al$_x$O$_y$-350[21] and Y$_x$O$_y$-350,[20] respectively. This phenomenon was also observed by Jeong *et al.* and was attributed to

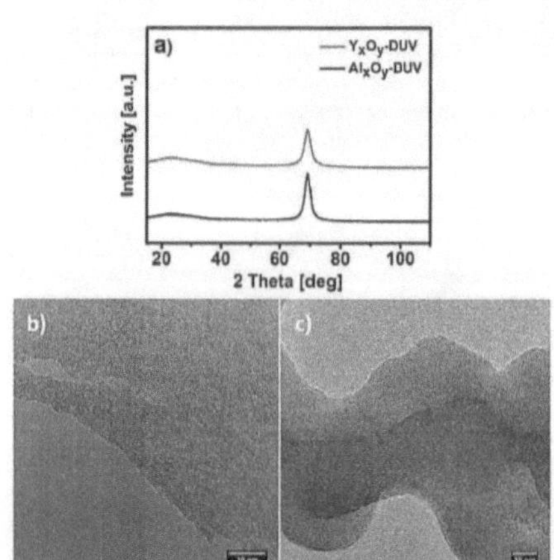

Fig. 3 (a) XRD patterns of DUV-processed Al$_x$O$_y$ and Y$_x$O$_y$ single phase dielectrics derived from precursors **1** and **2** and finally processed at 150 °C. TEM images of DUV-processed (b) Al$_x$O$_y$ and (c) Y$_x$O$_y$ dielectrics. The strong signal at 2 theta = 64° is due to the polycrystalline Si substrate.

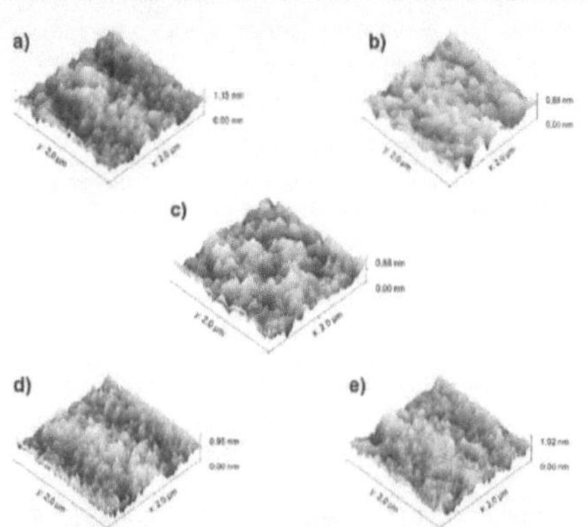

Fig. 5 AFM images of (a) 10-YAl$_x$O$_y$, (b) 30-YAl$_x$O$_y$ and (c) 50-YAl$_x$O$_y$ thin films processed at 350 °C and AFM images of DUV-processed (d) Al$_x$O$_y$ and (e) Y$_x$O$_y$ thin films at 150 °C.

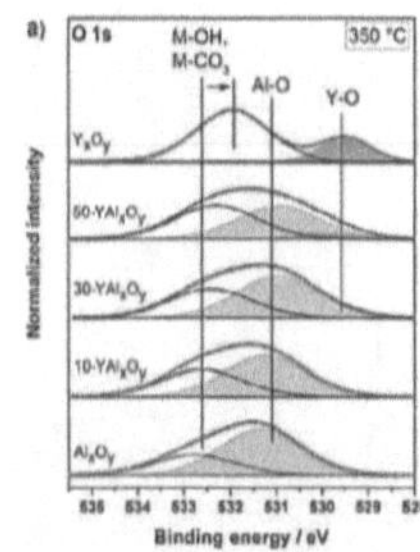

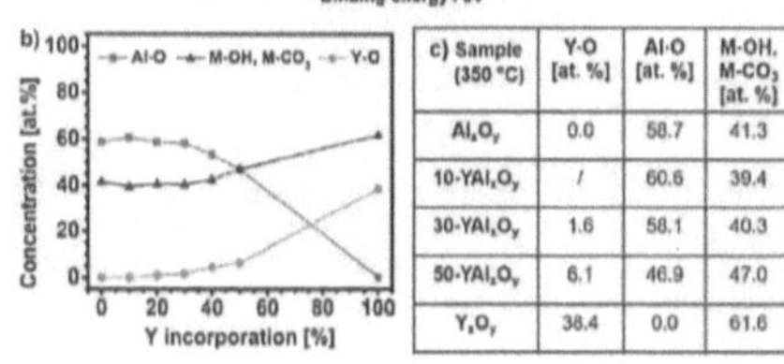

Fig. 6   O 1s XPS core spectra of YAl_xO_y thin films with various compositions processed at 350 °C in comparison to the binary Al_xO_y[21] and Y_xO_y[20] dielectrics.

c) Sample (350 °C)	Y-O [at. %]	Al-O [at. %]	M-OH, M-CO$_3$ [at. %]
Al_xO_y	0.0	58.7	41.3
10-YAl_xO_y	/	60.6	39.4
30-YAl_xO_y	1.6	58.1	40.3
50-YAl_xO_y	6.1	46.9	47.0
Y_xO_y	38.4	0.0	61.6

enhanced disorder within the Al_xO_y aluminum host lattice.[35] Such an extraordinary film smoothness was also observed for the DUV-processed thin films, exhibiting very smooth surfaces of 0.14 and 0.12 nm for Al_xO_y-DUV and Y_xO_y-DUV, respectively (Fig. 5d–e). Such a very smooth surface roughness is highly desirable since it might allow for a smooth dielectric/semiconductor interface formation in a TFT device. As a result, small interface trap densities and thus low leakage currents of the final devices might be realizable.

In order to understand the chemical environment of the various YAl_xO_y compositions, thin films with various Al/Y ratios (10–50 mol% Y) were obtained by processing solutions of **1** and **2** at an annealing temperature of 350 °C in an ambient atmosphere and investigating them by means of X-ray photoelectron spectroscopy (XPS, Fig. 6). The yttrium content of each sample was verified (Table 1). The O 1s core spectra of the various YAl_xO_y compositions (Fig. 6a) were deconvoluted into

two (10-YAl_xO_y) and three (30-YAl_xO_y, 50-YAl_xO_y) peaks, respectively, with binding energies at 529.7, 531.1 and around 532.4 eV. The peak at 529.7 eV is attributed to oxygen associated with O^{2-} coordinated to Y^{3+}[(Y–O)48–51] and the peak at 531.1 eV is attributed to oxygen associated with O^{2-} coordinated Al^{3+} (Al–O)[52,53] and represents the dominant peak for all YAl_xO_y compositions. No peak at the binding energy of 529.7 eV (Y–O) is detected for the 10-YAl_xO_y sample, which might however be due to a very small concentration of the Y–O species which lies below the XPS detection limit. The peak at around 532.4 eV corresponds to the hydroxide as well as carbonate species of aluminium and yttrium.[49,50] A clear differentiation between carbonate and hydroxide is precluded due to the fact that these species appear in the same range of binding energies.[54] It can be noted that, for a pure Al_xO_y sample, the binding energy of the metal carbonate/hydroxide peak is observed at 532.6 eV whereas the binding energy of Y–OH/Y–CO$_3$ appears at 532.0 eV. The peak stemming from the carbonate and hydroxide species for YAl_xO_y shifts as expected from 532.6 eV to 532.3 eV with a higher content of yttrium. The presence of carbonate species is proven by a characteristic peak at the binding energy of 290.0 eV in the C 1s core spectra.[50]

The intensities of the three peaks at 529.7 eV, 531.1 eV and 532.4 eV change with increasing yttrium incorporation, whereby the peak associated to Al–O decreases. The intensities of the peaks associated to Y–O as well as to M–OH and M–CO$_3$ increase when going from sample 10-YAl_xO_y to sample 50-YAl_xO_y (Fig. 6c). This intensity increase can be explained by an increasing Lewis acidic behaviour due to a higher Y^{3+} incorporation, which allows for a stronger carbonate formation (Table 1).

Regarding the DUV-processed Al_xO_y dielectric the O 1s core spectra (Fig. 7a) were deconvoluted into two peaks at binding energies of 531.3 and 532.6 eV. The peak at 531.3 eV corresponds

Table 1   Amounts of yttrium, carbonate and nitrogen species calculated on the basis of XPS quantitative analysis

	Yttrium amount $(Y/(Y + Al)) \times 100$ [at%]	–CO$_3$ (C 1s at 290 eV) [at%]	–NO$_2$, –NO$_3^-$ (N 1s at 403.3 eV and at 407.7 eV) [at%]
Al_xO_y-350 °C	—	2.6	—
10-YAl_xO_y-350 °C	7.7	1.8	—
30-YAl_xO_y-350 °C	25.0	2.6	—
50-YAl_xO_y-350 °C	41.0	3.1	—
Y_xO_y-350 °C	100.0	6.0	—
Al_xO_y-DUV	0.0	2.0	—
30-YAl_xO_y-DUV	25.9	4.5	0.6
Y_xO_y-DUV	100.0	7.4	1.3

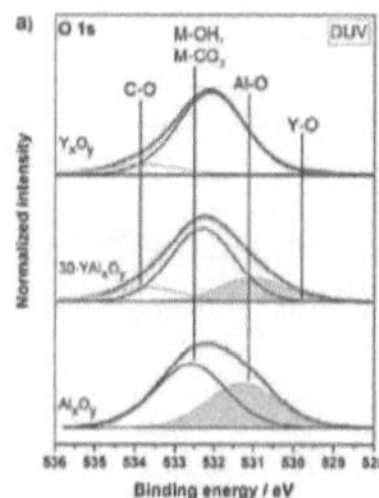

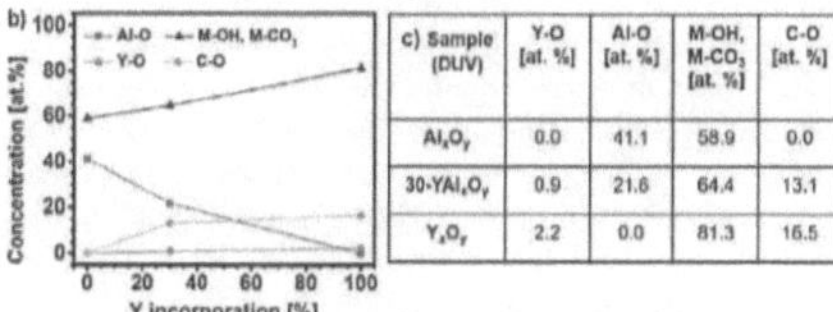

c) Sample (DUV)	Y-O [at. %]	Al-O [at. %]	M-OH, M-CO$_3$ [at. %]	C-O [at. %]
Al_xO_y	0.0	41.1	58.9	0.0
30-YAl_xO_y	0.9	21.6	64.4	13.1
Y_xO_y	2.2	0.0	81.3	16.5

Fig. 7   O 1s XPS core spectra of DUV-processed Al_xO_y, Y_xO_y and 30-YAl_xO_y dielectric thin films processed at 150 °C.

to oxygen associated with O^{2-} coordinated aluminium bonds (Al–O)[52,53] and the peak at 532.6 eV can be attributed to the hydroxide as well as carbonate species of aluminium (Al–OH, Al–CO$_3$).[55] The O 1s core spectra of the DUV-processed Y$_x$O$_y$ dielectric were deconvoluted into three peaks at binding energies of 529.5, 532.0 and 533.5 eV. Here, the peak at 529.5 eV can be attributed to oxygen associated with O^{2-} coordinated Y^{3+} (Y–O)[48–51] and the peak at 532.0 eV corresponds to the hydroxide as well as carbonate species of yttrium (Y–OH, Y–CO$_3$).[49,50] The signal at the binding energy of 533.5 eV refers to (C–O) fragments arising most probably due to an incomplete decomposition of the yttrium precursor **2** under the chosen conditions.[50] In accordance with this incomplete decomposition process, minor amounts of NO_3^- species are detected in the N 1s core spectrum, indicated by a signal at 407.1 eV[56,57] as well as –NO_2 groups at 403.3 eV.

The O 1s core spectra of the DUV-processed 30-YAl$_x$O$_y$ sample exhibits four peaks at binding energies of 529.5 (with very weak intensity), 531.3, 532.6 and 533.5 eV. Thereby the peak associated to Al–O (531.3 eV) decreases and the peak associated to the hydroxyl and carbonate species of aluminium and yttrium (532.6 eV) increases in comparison to the relevant species in the Al$_x$O$_y$ sample and can be explained by the presence of yttrium in the sample. Furthermore, in accordance with the pure Y$_x$O$_y$ sample the peak at 533.5 eV can be attributed to the C–O species. Furthermore, the 30-YAl$_x$O$_y$ sample shows signals which can be assigned to the presence of nitrogen containing groups due to the incomplete decomposition of the ligand species of precursor **2** under these conditions (Table 1).

Dielectric characterization

In order to determine the dielectric properties of the solution-processed metal oxide thin films, a metal–insulator–metal (MIM) structure (Fig. 8) was fabricated. For the fabrication of the capacitor devices, ITO-coated glass was used as the substrate, whereby the ITO layer serves as the gate electrode. The precursor solutions of **1**, **2** and various mixtures were spin-coated on the ITO-coated glass and either annealed at a temperature of 350 °C or treated with DUV irradiation leading to the formation of Al$_x$O$_y$, Y$_x$O$_y$ and YAl$_x$O$_y$ thin films, respectively. As the top electrode, a 60 nm gold layer was sputter-deposited onto the dielectric thin films.

Aluminium oxide-based capacitors with the incorporation of various yttrium amounts (10/30/50 mol%) were fabricated at a processing temperature of 350 °C (Fig. 9a and b), in order to investigate the influence of the dielectric properties. It is evident that the yttrium incorporation into the aluminium oxide based dielectric results in an incremental enhancement of the dielectric constant of the solution-processed YAl$_x$O$_y$ dielectrics (Fig. 9a). The dielectric constant of the solution-processed binary oxide Al$_x$O$_y$ dielectric is κ = 11.8.[21]

We could systematically increase the dielectric constant for the ternary oxide dielectrics up to κ = 12.0, 12.5, and 13.3 for 10-YAl$_x$O$_y$, 30-YAl$_x$O$_y$ and 50-YAl$_x$O$_y$, respectively. In addition, the frequency dispersion in the range from 10 to 100 kHz could

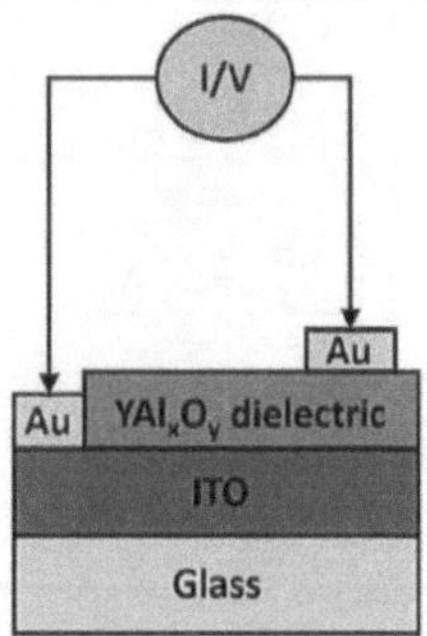

Fig. 8 Schematic illustration of the fabricated MIM-capacitor device. Grey, purple and blue layers represent the glass substrate, the 140 nm ITO film serving as the bottom gate electrode and the YAl$_x$O$_y$ dielectric thin film, respectively. The circular top electrodes and the sacrificial contact on the left-hand side are composed of 60 nm and 100 nm gold layers, respectively.

be successfully minimized by the incorporation of yttrium, whereby a significant reduction of the frequency dispersion is observable for the 30-YAl$_x$O$_y$ sample in comparison to the 10-YAl$_x$O$_y$ sample and the pristine Al$_x$O$_y$ sample.

Regarding the I–V characteristics at an electric field of 1 MV cm^{-1} (Fig. 9b), at first the incorporation of yttrium (10-YAl$_x$O$_y$) results in even lower leakage current densities in comparison to pristine Al$_x$O$_y$. It is apparent that a further steady incorporation of yttrium results in slightly higher leakage current densities, with 30-YAl$_x$O$_y$ slightly surpassing the value of the pristine Al$_x$O$_y$ dielectric. It is noteworthy that for all the investigated

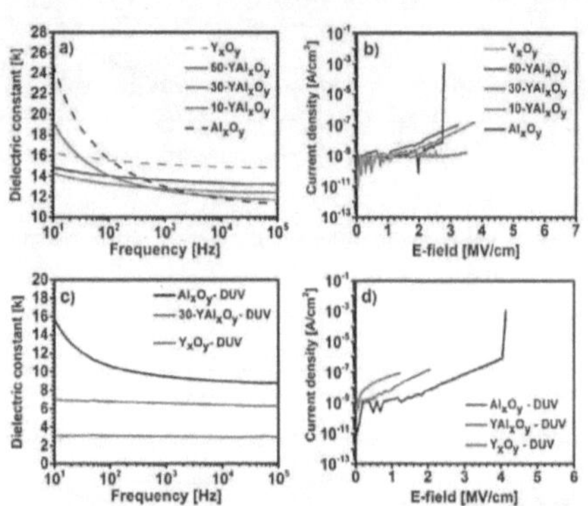

Fig. 9 (a) Dielectric constant vs. frequency curves and (b) the leakage current density behaviour of the solution-processed Al$_x$O$_y$, Y$_x$O$_y$ and YAl$_x$O$_y$ dielectrics with various compositions, annealed at 350 °C. (c) Dielectric constant vs. frequency curves and (d) leakage current density behaviour of solution-processed DUV-mediated Al$_x$O$_y$, Y$_x$O$_y$ and 30-YAl$_x$O$_y$ dielectrics processed at 150 °C.

compositions, the leakage current densities are in the same low order of magnitude ($J < 2 \times 10^{-9}$ A cm^{-2} at 1 MV cm^{-1}), fulfilling the criterion of Wager *et al.* for an ideal dielectric.[26] These observations are in full accord with the XPS studies (see Table 1), displaying the lowest amount of carbonate species for the 10-YAl$_x$O$_y$ sample. Further yttrium incorporation leads to an increased amount of carbonate species within the YAl$_x$O$_y$ samples, which corresponds to the increase of the leakage current densities for samples with higher yttrium proportions.

In summary, the use of the nitro functionalized malonato complexes of aluminium 1 and yttrium 2, in a ratio of 7:3 (30-YAl$_x$O$_y$), processed at 350 °C exhibits the best dielectric performance of the herein fabricated capacitors. The 30-YAl$_x$O$_y$-350 sample shows excellent dielectric properties and combines the desirable electrical attributes from both single metal oxide dielectrics (Al$_x$O$_y$/Y$_x$O$_y$) derived from the individual precursors 1 and 2. The 30-YAl$_x$O$_y$-350 sample exhibits an increased dielectric constant at 10 kHz of $\kappa = 12.5$ in comparison to $\kappa = 11.8$ for the Al$_x$O$_y$-350 sample. Furthermore, the 30-YAl$_x$O$_y$-350 sample impressively displays almost no frequency dispersion, which is one major drawback of the Al$_x$O$_y$-350[21] dielectric derived from precursor 1.

Regarding the *I–V* characteristics, the 30-YAl$_x$O$_y$-350 sample displays a slightly increased leakage current density in comparison to the Al$_x$O$_y$-350[21] dielectric, but also an increased breakdown voltage of $E_B > 3.7$ MV cm^{-1} compared to $E_B = 2.8$ MV cm^{-1} for the Al$_x$O$_y$-350[21] sample. Also noteworthy and very desirable is that the 30-YAl$_x$O$_y$ dielectric is amorphous throughout, while the known single phase dielectric Y$_x$O$_y$ has already started to crystallize at 350 °C,[20] which enhances the probability of short circuits when further down-scaling to thinner dielectric films. In comparison to the work of Jeong *et al.* the herein reported solution-processed YAl$_x$O$_y$ dielectrics utilizing precursors 1 and 2 lead to thin films with lower dielectric constants, but also lower leakage current densities and with no straightforward electrical breakdown up to 3.3 MV cm^{-1}, although processed at a slightly lower temperature of 350 °C in comparison to 400 °C (see Table 2). This represents a significant step further towards the development of solution processable high-κ dielectrics.

Another aim of this research was the investigation of an alternative processing route for the generation of thin film gate dielectrics, in order to reduce the PDA temperature, thus taking the materials a step further towards application in flexible electronics. Due to the strong absorbance of precursors 1 and 2 at wavelengths of $\lambda < 250$ nm (Fig. 1), we have chosen a treatment using DUV irradiation ($\lambda = 160$ nm) in order to generate active oxide layers therefrom. Fig. 9c and d display the dielectric properties of the low-temperature (150 °C) DUV-mediated and solution-processed Al$_x$O$_y$, Y$_x$O$_y$ and 30-YAl$_x$O$_y$ dielectrics. The DUV-processed Y$_x$O$_y$ and YAl$_x$O$_y$ dielectrics of all compositions exhibit electrical short circuits throughout while using a single layer of the yttrium and mixed precursor solutions. The subsequent utilization of three layers however results in an electrically well performing Y$_x$O$_y$ dielectric, nevertheless still possessing insufficient dielectric characteristics and exhibiting a very low dielectric constant of $\kappa = 2.9$ and a relatively high leakage current density of 5.8×10^{-8} A cm^{-2} at 1 MV cm^{-1}, especially when considering the very high film thickness of 329 nm. This observation might be due to an incomplete metal oxide formation paralleled by extensive void formation during the decomposition process. The Al$_x$O$_y$-DUV processed dielectric on the other hand shows excellent dielectric properties by using just a single layer of the aluminium precursor solution for the Al$_x$O$_y$ film formation. The DUV-processed Al$_x$O$_y$ sample exhibits a very high areal capacitance of 153 nF cm^{-2} ($\kappa = 9.0$) which can be attributed to the low film thickness of 52 nm ($C = \varepsilon_0 \cdot \kappa \cdot A/D$). Considering the low film thickness, the Al$_x$O$_y$-DUV dielectric possesses a remarkably low leakage current density at 1 MV cm^{-1} of just 1.7×10^{-9} A cm^{-2}. Furthermore, the Al$_x$O$_y$-DUV dielectric displays a very satisfactory breakdown voltage of 4.1 MV cm^{-1}. Taking into consideration that the DUV-solution-processed Y$_x$O$_y$ dielectric is supposed to be the material component with a high dielectric constant within the ternary YAl$_x$O$_y$ composition, the incorporation of yttrium into the Al$_x$O$_y$ lattice using the DUV-mediated approach should allow for an enhanced dielectric behaviour. However, unexpectedly, the attempted incorporation using precursor 2 deteriorates the dielectric performance of the pristine Al$_x$O$_y$ based capacitors. Fig. 9c and d show the deteriorated electrical properties of the DUV generated yttrium doped YAl$_x$O$_y$ compositions in comparison to the Al$_x$O$_y$-DUV dielectric.

Compared to the solution-processed Al$_x$O$_y$ and 30-YAl$_x$O$_y$ dielectrics reported by Bolat and Romanyuk *et al.* the herein fabricated Al$_x$O$_y$-DUV dielectric possesses a slightly higher dielectric constant (see Table 3). Regarding the leakage current

Table 2 Overview and comparison of the dielectric properties of solution-processed Al$_x$O$_y$, Y$_x$O$_y$ and YAl$_x$O$_y$ dielectrics

Ref.	Sample	Temp. [°C]	D [nm]	κ (10 kHz)	Freq. disp. of κ (10–100 kHz)	Leakage at 1 MV [A cm^{-2}]	E_B [MV cm^{-1}]
21	Al$_x$O$_y$	350	122	11.8	11.3–24.8	8.9×10^{-10}	2.8
This work	10-YAl$_x$O$_y$	350	114 ± 1.6	12.0 ± 1.4	11.7–19.2	$0.7(\pm 4.0) \times 10^{-9}$	>3.5
This work	30-YAl$_x$O$_y$	350	107 ± 1.2	12.5 ± 1.6	12.4–14.3	$1.1(\pm 2.3) \times 10^{-9}$	>3.7
This work	50-YAl$_x$O$_y$	350	118 ± 1.1	13.3 ± 1.3	13.2–14.8	$1.8(\pm 4.5) \times 10^{-9}$	>3.3
20	Y$_x$O$_y$	350	156	14.9	14.8–16.3	3.8×10^{-10}	>2.6
35	Al$_2$O$_3$	400	37	11.5	—	8.7×10^{-8}	~4.0
35	Al$_{1.52}$Y$_{0.48}$O$_3$	400	35	12.2	—	4.9×10^{-9}	~4.5
35	Al$_{1.02}$Y$_{0.92}$O$_3$	400	37	15.6	—	8.6×10^{-9}	~4.0
35	Al$_{0.45}$Y$_{1.55}$O$_3$	400	37	19.5	—	1.1×10^{-8}	~5.0
35	Y$_2$O$_3$	400	40	20.0	—	2.1×10^{-5}	~3.5

Table 3 Overview and comparison of the dielectric properties of DUV-mediated solution-processed Al_xO_y, Y_xO_y and YAl_xO_y dielectrics

Ref.	Sample (DUV)	D [nm]	κ (10 kHz)	Areal cap. [nF cm^{-2}]	Leakage at 1 MV [A cm^{-2}]	E_B [MV cm^{-1}]
This work	Y_xO_y	319 ± 13.2	2.9 ± 0.9	8 ± 2	$5.8(\pm 404.2) \times 10^{-8}$	>1.6
This work	$30\text{-}YAl_xO_y$	195 ± 8.6	6.4 ± 1.6	29 ± 3	$7.4(\pm 9.6) \times 10^{-9}$	>2.0
This work	Al_xO_y	52 ± 1.3	9.0 ± 0.7	153 ± 12	$1.7(\pm 1.6) \times 10^{-9}$	4.1
37	AlO_x	~ 60	~ 6.8	—	5.5×10^{-8}	>3.4
37	$30\text{-}YAlO_x$	~ 60	~ 8.9	—	1.9×10^{-8}	2.7

density at 1 MV cm^{-1} the 30-YAl$_x$O$_y$ dielectric, fabricated by Bolat and Romanyuk *et al.* and derived from the respective nitrate precursors, displays values that are one magnitude higher than the leakage current densities generated from the herein reported aluminium oxide derived from precursor 1, which is 1.7×10^{-9} A cm^{-2} (see Table 3). Furthermore, the herein reported Al$_x$O$_y$-DUV dielectric exhibits a higher breakdown voltage of 4.1 MV cm^{-1} in comparison to 2.7 and 3.4 MV cm^{-1} for the Al$_x$O$_y$ and 30-YAlO$_x$ dielectrics of the reference work, respectively (Table 3).

Functionality of the YAl$_x$O$_y$ dielectric in a field-effect transistor device

The feasibility of the solution-processed 30-YAl$_x$O$_y$-350 sample as a functional dielectric layer in a TFT device was evaluated. The fabrication of the TFT was achieved by using the 30-YAl$_x$O$_y$ dielectric and a solution-processed indium zinc oxide (IZO) semiconductor, which was subsequently annealed at 350 °C. The semiconducting IZO thin film used was fabricated using molecular single-source precursors of indium and zinc as described previously by our group.[20,21,41-43] Fig. 10 shows a schematic illustration of the TFT geometry.

The crucial TFT performance characteristics include the charge-carrier mobility (μ_{sat}) in the saturation regime, the on-voltage (V_{on}), the threshold voltage (V_{th}) and the ratio of the current in the on- and off-state ($I_{on/off}$) of the transistor device. They were extracted from the transfer and output curves (Fig. 11). The measured devices exhibit reasonable performance characteristics with a μ_{sat} value of 2.6 cm^2 V^{-1} s^{-1}, V_{on} of -1.1 V, V_{th} of 12.4 V and $I_{on/off}$ of 1.8×10^7, clearly demonstrating the functionality of the solution-processed 30-YAl$_x$O$_y$-350 dielectric when combined with an active IZO semiconductor in a TFT device. It is noteworthy that such solution-processed yttrium aluminium oxide dielectric of discrete composition Al$_{1.58}$Y$_{0.42}$O$_3$ when used as the gate dielectric enables

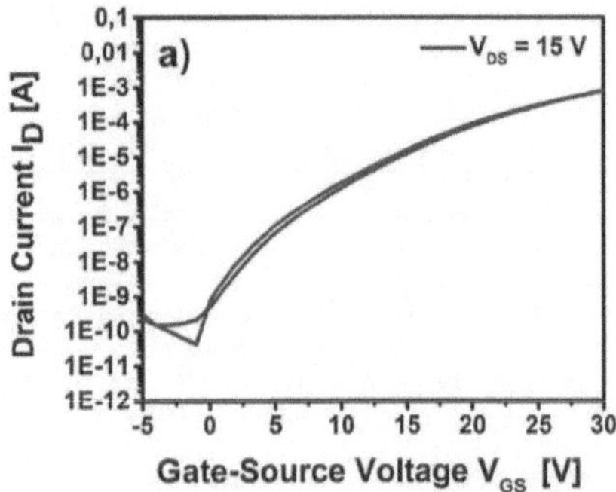

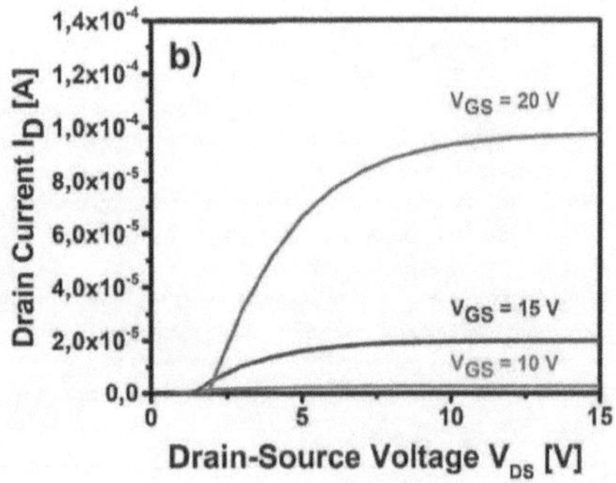

Fig. 11 Electrical characterization of a TFT device based on a solution-processed 30-YAl$_x$O$_y$ dielectric and a solution-processed indium zinc oxide (IZO) semiconductor, both processed at 350 °C. (a) Transfer characteristics of the device at V_{DS} = 15 V (forward and reverse sweep). (b) Output characteristics of the TFT device were measured at 10, 15 and 20 V, respectively.

the generation of even higher performing TFT devices (μ_{FE} = 47.7 cm^2 V^{-1} s^{-1} and $I_{on/off} \approx 10^6$). Herein, however, the semiconductor is DC-magnetron-sputtered which is eventually known to produce even higher mobilities.[35]

Conclusions

In summary, we have demonstrated the formation of a ternary yttrium aluminium oxide YAl$_x$O$_y$ dielectric utilizing the molecular single-source precursors tris[(diethyl-2-nitromalonato)]aluminium(III),

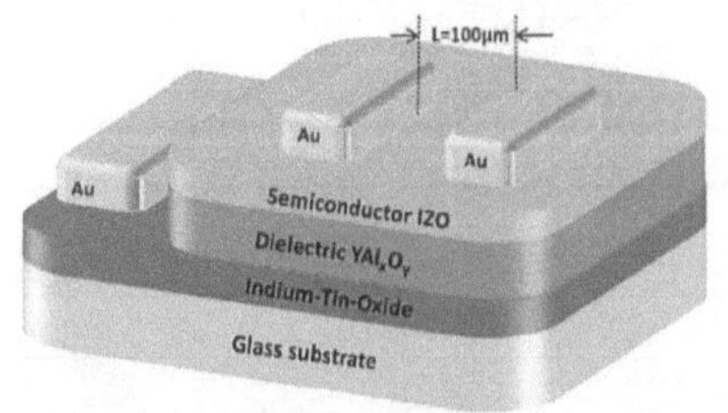

Fig. 10 Schematic illustration of the fabricated 30-YAl$_x$O$_y$ based thin-film transistor.

Al-DEM-NO$_2$ **1**, and bis(diethyl-2-nitromalonato) nitrato yttrium(III), Y-DEM-NO$_2$ **2**. Thereby, thermal (350 °C) as well as DUV-mediated low-temperature processing of metal oxide dielectrics from solution was performed due to the strong absorption of precursors at wavelengths of $\lambda < 250$ nm. The YAl$_x$O$_y$-350 samples of all the investigated compositions are amorphous throughout, even on the nanometer scale, and exhibit a very smooth surface roughness of $R_{RMS} < 0.2$ nm. The incorporation of yttrium leads to a non-linear behaviour of the dielectric film thickness. This non-linear behaviour can be attributed to the inclusion of yttrium into the Al$_x$O$_y$ host lattice, thus reaching saturation at an yttrium incorporation of $\sim$30 mol%. The XPS surface analysis reveals that the incorporation of yttrium into the amorphous Al$_x$O$_y$ lattice at first results in a decrease of the carbonate species within the material (10-YAl$_x$O$_y$, 30-YAl$_x$O$_y$), corresponding to a decreased leakage current density of the final capacitor device. Noteworthily, all YAl$_x$O$_y$-350 based capacitors exhibit a very low leakage current density of $J < 2 \times 10^{-9}$ A cm^{-2} at 1 MV cm^{-1}. In addition, no electrical breakdown occurs at voltages up to 40 V ($E_B > 3.3$ MV cm^{-1}) for all YAl$_x$O$_y$ compositions. Furthermore, the fabricated YAl$_x$O$_y$-350 based capacitors all possess increased dielectric constants of $\kappa = 12.0$, 12.5 and 13.3 at 10 kHz for 10-YAl$_x$O$_y$, 30-YAl$_x$O$_y$ and 50-YAl$_x$O$_y$, respectively. Besides, from an yttrium incorporation of 30 mol% onwards, almost no frequency dispersion over the range of 10 Hz–100 kHz can be observed.

Regarding the DUV-processed dielectrics, surprisingly the yttrium precursor **2** does not undergo the desirable photolytic decomposition at a wavelength of $\lambda = 160$ nm, resulting in poor YAl$_x$O$_y$ based dielectrics. This unexpected behaviour has to be part of future investigations. Nevertheless, the DUV-mediated Al$_x$O$_y$ dielectric itself displays excellent dielectric properties ($\kappa = 9.0$, $C_{Areal} = 153$ nF cm^{-2}, $J = 2 \times 10^{-9}$ A cm^{-2} at 1 MV cm^{-1} and $E_B = 4.1$ MV cm^{-1}), especially when considering a maximum processing temperature of 150 °C and using only a single layer deposition. Moreover, such DUV-mediated approaches can be extended towards the formation of other binary or multinary metal oxides utilizing the tailored precursor chemistry towards solution processing of oxide thin films. Such approaches can enable interesting functional properties such as direct photopatterning as well as low-temperature processing on flexible substrates or novel device applications in the future.

Conflicts of interest

The authors declare no conflicts of interest.

Acknowledgements

Transmission electron microscopy (TEM) investigations were performed under contract ERC-TUD1 at ERC Jülich, Germany. We would like to thank Dr J. Engstler and S. Heinschke for performing the TEM, spectroscopic ellipsometry and XRD studies.

References

1 R. P. Ortiz, A. Facchetti and T. J. Marks, *Chem. Rev.*, 2010, **110**, 205–239.

2 J. Robertson, *Rep. Prog. Phys.*, 2005, **69**, 327–396.

3 P. Barquinha, R. Martins, L. Pereira and E. Fortunato, *Transparent Oxide Electronics: From Materials to Devices*, 2012.

4 R. Martins, P. Barquinha, L. Pereira, I. Ferreira and E. Fortunato, *Appl. Phys. A: Mater. Sci. Process.*, 2007, **89**, 37–42.

5 Y.-J. Cho, J.-H. Shin, S. Bobade, Y.-B. Kim and D.-K. Choi, *Thin Solid Films*, 2009, **517**, 4115–4118.

6 Y. S. Chun, S. Chang and S. Y. Lee, *Microelectron. Eng.*, 2011, **88**, 1590–1593.

7 P. Barquinha, L. Pereira, G. Gonçalves, R. Martins, E. Fortunato, D. Kuscer, M. Kosec, A. Vilà, A. Olziersky and J. R. Morante, *J. Soc. Inf. Disp.*, 2010, **18**, 762–772.

8 E. Carlos, J. Leppäniemi, A. Sneck, A. Alastalo, J. Deuermeier, R. Branquinho, R. Martins and E. Fortunato, *Adv. Electron. Mater.*, 2020, **6**, 1901071.

9 W. Xu, H. Wang, L. Ye and J. Xu, *J. Mater. Chem. C*, 2014, **2**, 5389–5396.

10 C.-G. Lee and A. Dodabalapur, *J. Electron. Mater.*, 2012, **41**, 895–898.

11 A. Liu, G. X. Liu, H. H. Zhu, F. Xu, E. Fortunato, R. Martins and F. K. Shan, *ACS Appl. Mater. Interfaces*, 2014, **6**, 17364–17369.

12 L. Zhu, G. He, J. Lv, E. Fortunato and R. Martins, *RSC Adv.*, 2018, **8**, 16788–16799.

13 Y. B. Yoo, J. H. Park, K. H. Lee, H. W. Lee, K. M. Song, S. J. Lee and H. K. Baik, *J. Mater. Chem. C*, 2013, **1**, 1651–1658.

14 R. Frunza, B. Kmet, M. Jankovec, M. Topic and B. Malic, *Mater. Res. Bull.*, 2014, **50**, 323–328.

15 G. Adamopoulos, S. Thomas, D. D. C. Bradley, M. A. McLachlan and T. D. Anthopoulos, *Appl. Phys. Lett.*, 2011, **98**, 123503.

16 K. Song, W. Yang, Y. Jung, S. Jeong and J. Moon, *J. Mater. Chem.*, 2012, **22**, 21265–21271.

17 F. Xu, A. Liu, G. Liu, B. Shin and F. Shan, *Ceram. Int.*, 2015, **41**, S337–S343.

18 A. Liu, G. Liu, H. Zhu, Y. Meng, H. Song, B. Shin, E. Fortunato, R. Martins and F. Shan, *Curr. Appl. Phys.*, 2015, **15**, S75–S81.

19 C.-Y. Tsay, C.-H. Cheng and Y.-W. Wang, *Ceram. Int.*, 2012, **38**, 1677–1682.

20 N. Koslowski, R. C. Hoffmann, V. Trouillet, M. Bruns, S. Foro and J. J. Schneider, *RSC Adv.*, 2019, **9**, 31386–31397.

21 N. Koslowski, S. Sanctis, R. C. Hoffmann, M. Bruns and J. J. Schneider, *J. Mater. Chem. C*, 2019, **7**, 1048–1056.

22 C. Avis and J. Jang, *J. Mater. Chem.*, 2011, **21**, 10649–10652.

23 W. Xu, H. Wang, F. Xie, J. Chen, H. Cao and J.-B. Xu, *ACS Appl. Mater. Interfaces*, 2015, **7**, 5803–5810.

24 A. Liu, G. Liu, H. Zhu, B. Shin, E. Fortunato, R. Martins and F. Shan, *RSC Adv.*, 2015, **5**, 86606–86613.

25 P. K. Nayak, M. N. Hedhili, D. Cha and H. N. Alshareef, *Appl. Phys. Lett.*, 2013, **103**, 033518.

26 J. F. Wager, D. A. Keszler and R. E. Presley, *Transparent Electronics*, Springer, New York, 2008.

27 P. N. Plassmeyer, K. Archila, J. F. Wager and C. J. Page, *ACS Appl. Mater. Interfaces*, 2015, **7**, 1678–1684.

28 W. Yang, K. Song, Y. Jung, S. Jeong and J. Moon, *J. Mater. Chem. C*, 2013, **1**, 4275–4282.

29 I. Krylov, D. Ritter and M. Eizenberg, *J. Appl. Phys.*, 2017, **122**, 034505.

30 Y. N. Gao, Y. L. Xu, J. G. Lu, J. H. Zhang and X. F. Li, *J. Mater. Chem. C*, 2015, **3**, 11497–11504.

31 S. T. Meyers, J. T. Anderson, D. Hong, C. M. Hung, J. F. Wager and D. A. Keszler, *Chem. Mater.*, 2007, **19**, 4023–4029.

32 B. N. Pal, B. M. Dhar, K. C. See and H. E. Katz, *Nat. Mater.*, 2009, **8**, 898–903.

33 G. Teowee, K. C. McCarthy, F. S. McCarthy, T. J. Bukowski, D. G. Davis and D. R. Uhlmann, *J. Sol-Gel Sci. Technol.*, 1998, **13**, 895–898.

34 X. Huang and P. Lai, *IEEE International Conference of Electron Devices and Solid-State Circuits (EDSSC)*, 2010, pp. 1–4.

35 J. Lee, H. Seul and J. K. Jeong, *J. Alloys Compd.*, 2018, **741**, 1021–1029.

36 W. J. Scheideler, M. W. McPhail, R. Kumar, J. Smith and V. Subramanian, *ACS Appl. Mater. Interfaces*, 2018, **10**, 37277–37286.

37 S. Bolat, P. Fuchs, S. Knobelspies, O. Temel, G. T. Sevilla, E. Gilshtein, C. Andres, I. Shorubalko, Y. Liu and G. Tröster, *Adv. Electron. Mater.*, 2019, **5**, 1800843.

38 Y.-H. Kim, J.-S. Heo, T.-H. Kim, S. Park, M.-H. Yoon, J. Kim, M. S. Oh, G.-R. Yi, Y.-Y. Noh and S. K. Park, *Nature*, 2012, **489**, 128–132.

39 S. Sanctis, R. C. Hoffmann, R. Precht, W. Anwand and J. J. Schneider, *J. Mater. Chem. C*, 2016, **4**, 10935–10944.

40 A. H. Adl, P. Kar, S. Farsinezhad, H. Sharma and K. Shankar, *RSC Adv.*, 2015, **5**, 87007–87018.

41 S. Sanctis, N. Koslowski, R. Hoffmann, C. Guhl, E. Erdem, S. Weber and J. R. J. Schneider, *ACS Appl. Mater. Interfaces*, 2017, **9**, 21328–21337.

42 S. Sanctis, R. C. Hoffmann, M. Bruns and J. J. Schneider, *Adv. Mater. Interfaces*, 2018, **5**, 1800324.

43 R. C. Hoffmann, M. Kaloumenos, S. Heinschke, E. Erdem, P. Jakes, R.-A. Eichel and J. J. Schneider, *J. Mater. Chem. C*, 2013, **1**, 2577–2584.

44 S. Sanctis, R. C. Hoffmann and J. J. Schneider, *RSC Adv.*, 2013, **3**, 20071–20076.

45 S. Sanctis, R. C. Hoffmann, S. Eiben and J. J. Schneider, *Beilstein J. Nanotechnol.*, 2015, **6**, 785–791.

46 S. Sanctis, *Multinary metal oxide semiconductors – A study of different material systems and their application in thin-film transistors*, Technische Universität, 2020.

47 J. H. Scofield, *J. Electron Spectrosc. Relat. Phenom.*, 1976, **8**, 129–137.

48 D. Barreca, G. A. Battiston, D. Berto, R. Gerbasi and E. Tondello, *Surf. Sci. Spectra*, 2001, **8**, 234–239.

49 C. Durand, C. Dubourdieu, C. Vallée, V. Loup, M. Bonvalot, O. Joubert, H. Roussel and O. Renault, *J. Appl. Phys.*, 2004, **96**, 1719–1729.

50 T. Gougousi and Z. Chen, *Thin Solid Films*, 2008, **516**, 6197–6204.

51 E. J. Rubio, V. V. Atuchin, V. N. Kruchinin, L. D. Pokrovsky, I. P. Prosvirin and C. V. Ramana, *J. Phys. Chem. C*, 2014, **118**, 13644–13651.

52 B. Strohmeier, *Surf. Sci. Spectra*, 1994, **3**, 135–140.

53 V. Trouillet, H. Troesse, M. Bruns, E. Nold and R. White, *J. Vac. Sci. Technol., A*, 2007, **25**, 927–931.

54 J. Stoch and J. Gablankowska-Kukucz, *Surf. Interface Anal.*, 1991, **17**, 165–167.

55 C. Gao, X.-Y. Yu, R.-X. Xu, J.-H. Liu and X.-J. Huang, *ACS Appl. Mater. Interfaces*, 2012, **4**, 4672–4682.

56 M. Y. Smirnov, A. Kalinkin and V. Bukhtiyarov, *J. Struct. Chem.*, 2007, **48**, 1053–1060.

57 B. C. Beard, *Appl. Surf. Sci.*, 1990, **45**, 221–227.

6.4 Toward an Understanding of Thin-Film Transistor Performance in Solution-Processed Amorphous Zinc Tin Oxide (ZTO) Thin Films

Zinc tin oxide (ZTO) represents a promising candidate for indium-free, amorphous oxide semiconductors still possessing comparable performance characteristics to the indium-based semiconductors. Up to date, various zinc salts (nitrates, chlorides, acetates) have been extensively investigated as potential zinc precursor in ZTO, while the choice of the tin precursor has been mainly restricted to tin (II) chloride so far. As a consequence, higher thermal decomposition temperatures are required. Furthermore, the use of chlorides can deteriorate the electrical performance of the final device due to the incorporation of Cl$^-$ ion trace impurities as well as an increased porosity of the thin films due to the evolution of acidic byproducts like HCl.

Herein, we demonstrate the synthesis and full structural characterization of the new tin (II) precursor coordination compound bis[(methoxyimino)-propanoato]tin(II) (tin-oximato), containing methoxyiminopropionic acid ligands, analogues to the previously established zinc precursor. The employment of both precursors allows the formation of functional ZTO semiconductor thin films with good electrical performance characteristics, when it is implemented in a TFT device. Furthermore, the combination of the zinc and tin (II) precursors results in a homogeneous start of the thermal decomposition in a narrow temperature range between 125 and 150 °C, due to a similar thermal decomposition behavior, verified by thermogravimetric analysis (TGA). X-ray diffraction (XRD) analysis reveals that the decomposition of the tin (II) precursor already exhibits crystalline reflexes at temperatures as low as 350°C, while the ZTO sample retains the amorphous state. XPS analysis confirms the SnO_2 chemical state in the amorphous phase, as well as the absence of the SnO species in the ZTO material. Additionally, no significant contamination of residual hydrocarbon in the final ZTO ceramic can be detected. Auger electron spectroscopy was performed to confirm a uniform distribution of the Zn, Sn(IV) and O species across the surface as well as in the depth of the films.

Furthermore, it was possible to determine and compare the defect states of SnO_2 and ZTO derived from the respective oximato precursors via electron paramagnetic resonance (EPR) spectroscopy. According to the core–shell model, SnO_2 exhibits a higher defect concentration in comparison to ZTO, exhibiting EPR defect states with a resonance at $g_{iso} = 2.0040$ for SnO_2 and $g_{iso} = 2.0032$ for the ZTO, which can be mainly attributed to surface defect spin states. As a result, a precursor ratio of 7:3 of Sn:Zn delivers optimum TFT performance characteristics of the ZTO-based TFT devices with a μ_{sat} of 5.18 cm^2 V^{-1}s^{-1}, V_{th} of 7.5 and $I_{on/off}$ of 6 x 10^8, employing a PDA temperature of 350 °C.

Toward an Understanding of Thin-Film Transistor Performance in Solution-Processed Amorphous Zinc Tin Oxide (ZTO) Thin Films

Shawn Sanctis,[†] Nico Koslowski,[†] Rudolf Hoffmann,[†] Conrad Guhl,[‡] Emre Erdem,[§] Stefan Weber,[§] and Jörg J. Schneider*,[†]

[†]Fachbereich Chemie, Eduard-Zintl-Institut, Fachgebiet Anorganische Chemie, Technische Universität Darmstadt, Alarich-Weiss-Straße 12, 64287 Darmstadt, Germany

[‡]Fachgebiet Surface Science, Technische Universität Darmstadt, Jovanka-Bontschits-Straße 2, 64287 Darmstadt, Germany

[§]Institute of Physical Chemistry, Universität Freiburg, Albert Straße 21, 79104 Freiburg, Germany

ⓢ *Supporting Information*

ABSTRACT: Amorphous zinc tin oxide (ZTO) thin films are accessible by a molecular precursor approach using mononuclear zinc(II) and tin(II) compounds with methoxyiminopropionic acid ligands. Solution processing of two precursor solutions containing a mixture of zinc and tin(II)−methoxyiminopropinato complexes results in the formation of smooth homogeneous thin films, which upon calcination are converted into the desired semiconducting amorphous ZTO thin films. ZTO films integrated within a field-effect transistor (FET) device exhibit an active semiconducting behavior in the temperature range between 250 and 400 °C, giving an

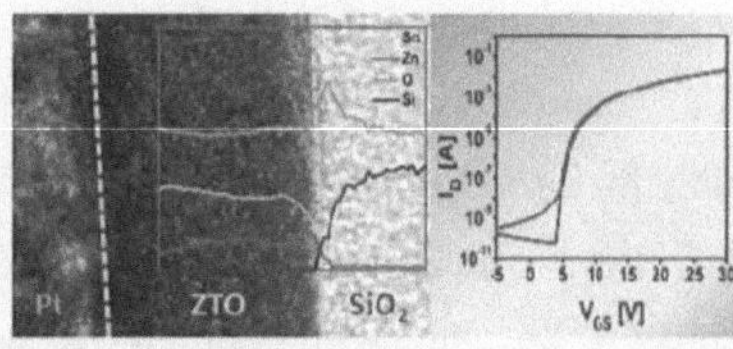

increased performance, with mobility values between $\mu = 0.03$ and 5.5 cm²/V s, with on/off ratios increasing from 10^5 to 10^8 when going from 250 to 400 °C. Herein, our main emphasis, however, was on an improved understanding of the material transformation pathway from weak to high performance of the semiconductor in a solution-processed FET as a function of the processing temperature. We have correlated this with the chemical composition and defects states within the microstructure of the obtained ZTO thin film via photoelectron spectroscopy (X-ray photoelectron spectroscopy, ultraviolet photoelectron spectroscopy), Auger electron spectroscopy, electron paramagnetic resonance spectroscopy, atomic force microscopy, and photoluminescence investigations. The critical factor observed for the improved performance within this ZTO material could be attributed to a higher tin concentration, wherein the contributions of point defects arising from the tin oxide within the final amorphous ZTO material play the dominant role in governing the transistor performance.

KEYWORDS: *solution process, metal oxide, zinc tin oxide, thin-film transistors, EPR, field-effect transistor, semiconductor*

1. INTRODUCTION

Transparent amorphous metal oxide-based thin-film transistors (TFTs) have received significant attention since their inception by Hosono et al.[1,2] Although oxides, namely, indium zinc oxide (IZO) and indium gallium zinc oxide, have been the preferred compositional choice of the semiconductor, indium-free alternatives have also received significant attention due the estimated scarcity of naturally available indium resources.[3,4] Zinc tin oxide (ZTO), among others, has emerged as a promising candidate for indium-free semiconductors, enabling competitive device performance in comparison with its indium-based counterparts.[5−7]

In a more general manner, solution processing of such amorphous oxide semiconductors paves the pathway for future lower-cost applications, still exhibiting performances greater than those of conventional silicon in terms of their crucial performance parameters, such as charge-carrier mobility and current on−off ratio.[8−11] Although a wide variety of zinc salts (acetates, nitrates, and chlorides) have been investigated as possible precursors for zinc in ZTO, with regard to the effect on the final field-effect transistor (FET) performance, analogous alternatives to the starting tin precursor have largely been restricted to tin(II) chloride.[4,12] Although FETs based on the aforementioned precursor have shown good applicability, they seem to suffer from the requirement of a higher thermal decomposition temperature as well as an increased degree of porosity, due to the evolution of acidic byproducts like HCl vapors during the calcinations of the precursor thin films as well as residual Cl⁻ ion traces present within the final ceramic.[13−16] Such undesired effects can have a significant impact on the overall performance of the final device. Recent strategies via a redox chloride elimination reaction have been successfully demonstrated, which enabled high device performance based on solution-processed amorphous ZTO-based devices.[17]

Received: May 3, 2017
Accepted: June 2, 2017
Published: June 2, 2017

◆ ACS Publications   © 2017 American Chemical Society   **21328**   DOI: 10.1021/acsami.7b06203
ACS Appl. Mater. Interfaces 2017, 9, 21328−21337

In recent works, we have demonstrated the use of molecular precursors, especially coordination compounds containing specifically tailored ligands, that are prone to controlled decomposition and release of products of this process. The use of these compounds also enables application of a precursor–solvent system, which allows solution processing of the resulting metal oxide thin films, without the use of any additives or stabilizers, which could have an ambiguous impact on the formation of the resulting films.[18−20] Herein, we report on the use of a new tin (II) precursor complex, which along with the previously established zinc precursor, both containing methoxyiminopropionic acid-based ligands, allows the formation of a functional ZTO semiconductor with decent-to-very good performance in FET devices.[21] The functional properties of the obtained FETs have been optimized with respect to temperature and their resulting performance could be correlated with the degree of oxidation and formation of defects within the final functional ZTO thin film-based FETs.

2. EXPERIMENTAL SECTION

2.1. Precursor Synthesis. Bis(methoxyiminopropanoato)tin(II) (**1**): The reaction was carried out under inert gas conditions and with dried solvents. Anhydrous tin(II) chloride (1.89 g; 10 mmol) was dissolved in tetrahydrofuran (THF, 100 mL) at room temperature. Thereafter, ammonium 2-(methoxyimino)propionate (2.70 g, 20 mmol), which was obtained from the reaction of methoxyiminopropionic acid and ammonium hydrogen carbonate,[22] was added to this solution. Subsequently, the suspension was heated to 65 °C and stirred for 18 h. The reaction mixture was filtered, and the solvent, removed under vacuum. Evaporation of the solvent yielded a pale white solid product, which was dissolved in dichloromethane (35 mL), filtered, and precipitated again with methyl *tert*-butylether (210 mL). The air-stable final product (2.49 g, 71.05 % of the theoretical yield) crystallized in the form of thin fibrous needles after heating in THF to 60 °C followed by cooling in a refrigerator to 4 °C. Ceramic yield (CY) and elemental analysis (CHN): found CY 42.3 %, C 26.90 %, H 3.96 %, N 9.78 %. Calcd for $Sn(C_4H_6O_3N)_2$: CY 42.9 %, C 27.38 %, H 3.44 %, N 7.98 %. ^1H NMR (500 MHz, chloroform-d_1, 25 °C): δ = 4.02 ($-OCH_3$); 2.00 ($-N{=}C{-}CH_3$) ppm. $^{13}C(^1H)$ NMR (500 MHz, chloroform-d_1, 25 °C): δ = 168.46 ($-\underline{C}OO$); 152.35 ($-N{=}\underline{C}-$); 62.70 ($-O-\underline{C}H_3$); 11.48 ($-N{=}C-\underline{C}H_3$) ppm. Electron ionization mass spectrometry m/z 31 (36%), 40 (54%), 41 (99%), 44 (100%), 72 (39%), 117 (17%) 151 (15%), 190 (8%), 236 (20%), 253 (M^+).

Bis(methoxyiminopropanoato)zinc(II) (**2**): Solid diaqua-bis-(methoxyiminopropanoato)zinc(II)[23] was heated at 75 °C under vacuum (10^{-3} Torr) for 3 h to obtain **2**. No color change was observed.

2.2. FET Characterization. FET substrates were obtained from Fraunhofer IMPS, Dresden. The substrates (15×15 mm^2) consist of highly n-doped silicon with a 90 nm silicon oxide dielectric layer. The source–drain electrodes, with channel length $L = 20$ μm and channel width $W = 10$ mm ($W/L = 500$), were fabricated with a 40 nm gold interdigital structure and a 10 nm intermediate adhesion layer of indium tin oxide. Quartz substrates (15×15 mm^2) were used for the deposition of the precursor films, which were carefully scratched off; the obtained powders were used for X-ray diffraction (XRD) investigations. All substrates were sequentially cleaned with acetone, water, and 2-propanol for 10 min, respectively. The cleaned substrates were exposed for 2 min to oxygen plasma prior to spin coating to improve adhesion of the spin-coating solutions.

For the preparation of the spin-coating solutions, stock solutions of tin(II) and the zinc precursor were prepared by dissolving 1 wt % of the respective precursors in 2-propanol. After complete dissolution, a clear solution was obtained in both cases. The solutions were then filtered through a 0.22 μm PTFE syringe filter and mixed in the desired Sn/Zn ratio for spin coating the ZTO thin films. Optimized

precursor ratios of Sn/Zn were achieved by fabricating devices with different ratios. A relatively lower Sn concentration resulted in poor device performance, whereas an extremely high Sn concentration led to a conductive behavior (Figure S7). On the basis of the obtained results, an optimized Sn/Zn molar ratio of 7:3 was used for all ZTO films produced, which was directly used for spin-coating and fabrication of thin films. The spin coating was carried out under ambient conditions. Thin films were produced with a spinning speed of 2500 rpm for 20 s. The final semiconducting tin oxide thin films were obtained from these spin-coated solutions by thermal treatment on a hot plate at different temperatures for 2 h.

2.3. Material Characterization. Transmission electron microscopy (TEM) was carried out using Tecnai F20 (FEI), with an operating voltage of 200 kV. Samples for the TEM investigations were prepared by depositing diluted precursor solutions onto SiO$_2$-coated gold grids and annealing at desired temperatures. Atomic force microscopy (AFM) measurements were carried out with CP-II (Bruker-Veeco), with silicon cantilevers. Thermogravimetric analysis (TGA) measurements were performed using a TG209F1-Iris (Netzsch) thermal analyzer, employing aluminum crucibles. XRD was performed on a Rigaku Miniflex 600@40 kV 15 mA diffractometer using Cu Kα_1 radiation ($\lambda = 1.541$ Å). XPS measurements were carried out in an integrated ultra-high-vacuum system with a base pressure of 1×10^{-9} mbar, equipped with a PHI 5000 VersaProbe photoelectron spectrometer. XPS measurements were performed using a mono-chromatized Al Kα X-ray source. The hemispherical photoelectron analyzer was operated with a constant analyzer energy of 10 eV (pass energy) to obtain the core spectra with an energy resolution of about 0.6 eV. The spectrometer was calibrated to the Fermi edge and core level lines of sputtered copper foils. Valence band measurements were made using ultraviolet photoelectron spectroscopy (UPS), employing an excitation energy of He I (21.2 eV) and a negative bias of 6 V. The work function (WF) was determined by the secondary electron cutoff. The obtained spectra were fitted with Voigt functions and a Shirley background. X-band (9.86 GHz) electron paramagnetic resonance (EPR) measurements were performed with a Bruker EMX spectrometer using a rectangular TE102 (X-band) resonator. The magnetic field was determined using an NMR gaussmeter (ER 035M; Bruker); for magnetic-field calibration, polycrystalline DPPH with g = 2.0036 was used. Photoluminescence (PL) measurements were performed on Si substrates with an excitation wavelength of 250 nm. Auger spectroscopy: PHI 680 (Physical Electronics) scanning Auger nanoprobe operated at an acceleration voltage of 20 keV and a current of 10 nA. Sputtering was carried out under UHV (5×10^{-9} Torr), with an argon ion gun operated at 250 eV and 500 nA.

3. RESULTS AND DISCUSSION

Precursor **1**, $Sn(C_4H_6O_3N)_2$, could be synthesized by the metathesis reaction using tin(II) chloride and ammonium methoxyimino propionate in THF (Scheme 1). The reaction has to proceed at an elevated temperature to ensure complete conversion of the educts into the desired product. At room temperature only, mixtures of **1** and ammonium salts were formed, which could not be further purified. In **1**, the oximato ligand acts a chelating ligand with bidentate N,O coordination toward Sn(II), as reported for polynuclear Sn(IV) complexes with cyano-substituted oximates.[24] Coordination of Sn(II) to the carboxylate groups with solely an O,O coordination of the Schiff base ligand, resembling the bonding in tin(II) acetate, would lead to a different coordination environment with inconsistent spectroscopic data.[25] The thermal decomposition of **1** in oxygen as well as in argon was studied (Figure S1a). The residual mass of 42.31 % corresponds to the quantitative formation of tin(IV) oxide SnO$_2$ in oxygen, with an expected CY of 42.98%. The decomposition pathway of **1** under argon, however, leads to a different decomposition behavior, resulting in metallic tin and tin(IV)oxide as final products, probably via

DOI: 10.1021/acsami.7b06203
ACS Appl. Mater. Interfaces 2017, 9, 21328−21337

Scheme 1. Reaction Scheme for the Synthesis and Thermal Decomposition of 1 to $Sn(IV)O_2$ in Air

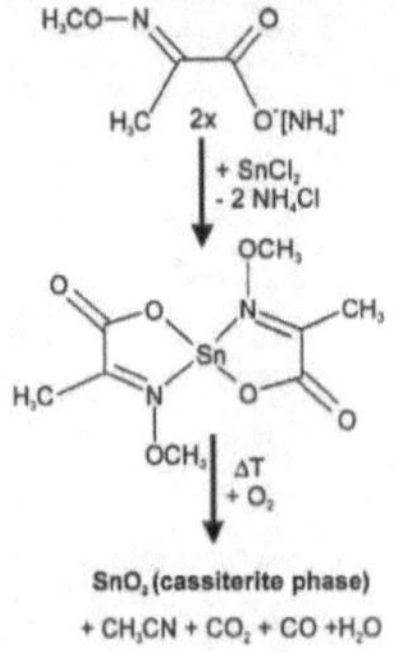

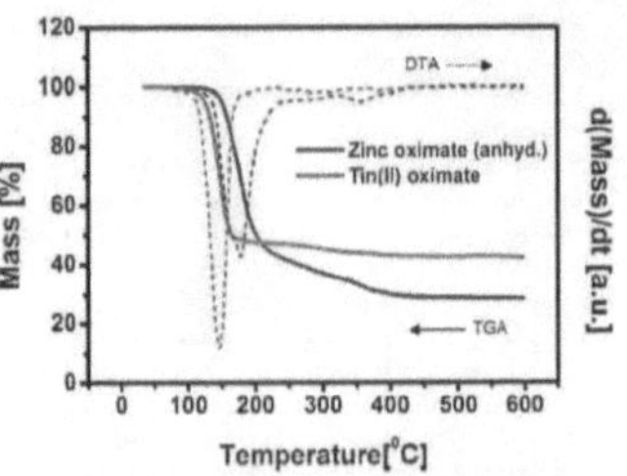

Figure 1. Thermogravimetric mass loss (solid) and differential thermal analysis (dotted) of zinc and tin(II) precursors 1 and 2 under oxygen.

the formation of subvalent $Sn(II)O$ as an intermediate. However, no evidence (see detailed XPS study below) for the formation of metallic tin during decomposition of 1 under oxygen was found, as, for example, reported during the thermal decomposition of tin(II)alkoxides.[26] This qualifies 1 as a clean and versatile new precursor for tin(IV) oxide. The thermal decay under oxygen (Scheme 1) proceeds in a single step at about 125 °C. The gaseous products, as analyzed by means of on-line mass spectrometry, are in agreement with a decomposition process according to a second-order Beckmann reaction (Figure S1b).[23] Thus, acetonitrile (m/z^+ 41) could be observed apart from more unspecific products such as water (m/z^+ 18), carbon dioxide (m/z^+ 44), and carbon monoxide (m/z^+ 28).

The combination of a solution of 1 with the established zinc(II) oximato complex diaqua-bis-(methoxyiminopropanoato)zinc(II),[23] which was successfully used so far for the formation of semiconducting ZnO and amorphous IZO, was found to be disadvantageous in the formation of ZTO using 1 as the molecular tin precursor. This is due to a subsequent hydrolytic cleavage of 1 by the coordinated aquo ligand of the diaqua-bis-(methoxyiminopropanoato)zinc(II) precursor. However, employment of the water-free bis(methoxyiminopropanoato)zinc-(II) compound (2) was found to help overcome this problem. Precursor 2 $[Zn(C_4H_6O_3N)_2]$ could be obtained from 1 by heating under vacuum[23] (Figure S2 and S3). Combining 1 and 2 allows a homogeneous start of the thermal decay in a narrow window between 125 and 150 °C due to the very similar thermal decomposition behavior (Figure 1). However, a somewhat more sluggish mass loss upon further heating to a higher temperature until the theoretical CY for the zinc precursor 2 compared to that for the tin counterpart 1 is observed.

The products of thermal transformation of 1 as well as of mixtures of 1 and 2 into the oxidic phases were investigated by means of high-resolution TEM (HRTEM) (Figure 2). Therefore, precursor solutions of 1 for SnO_2 and 1 and 2 (ratio 7:3) for ZTO were directly calcined on a TEM grid. In addition, a cross-section of a ZTO film of ~6 nm, prepared from the precursor solutions on a SiO_2/Si substrate and

encapsulated with a protective platinum layer, was manufactured using a focused ion beam (FIB). The precursor solutions of 1 and 2 afforded the desired amorphous ZTO phase after calcination, whereas the tin(II) precursor solution containing 1 after deposition and final calcination forms nanocrystalline tin oxide particles.

Moreover film formation at the interface of the SiO_2 layer of the silicon substrate is very smooth and shows no void formation at the dielectric SiO_x interface. XRD analysis of these calcined thin films between 350 and 400 °C, confirmed that tin(IV) oxide, SnO_2 in its tetragonal cassiterite phase (JCPDS No. 77-0450), was obtained. The broadening of the diffraction peaks in the powder diffractogram corroborates the finding of nanoparticulate SnO_2 obtained from the TEM results (Figure S4). Peak broadening could be observed at lower processing temperatures only, potentially indicative of its amorphous nature in those temperature regimes. No other contributions were found in the X-ray diffractograms. Auger electron spectroscopy contributing a surface as well as depth profile was performed for a ZTO thin film annealed at 350 °C to shed light on the distribution of the compositional elements within the final ZTO thin film obtained after spin coating the precursor solution of 1 and 2, followed by final calcination. A homogeneous distribution of the individual components of Zn, Sn, and O was obtained, confirming the desired chemical composition across the thin film (Figure 3). AFM studies of the SnO_2 and ZTO thin films revealed a homogeneous film surface and substrate coverage, with RMS values well below 1 nm for both (Figure 4). ZTO showed a relatively smoother surface at 350 °C with respect to its 250 °C counterpart, with a reduction in the surface texture, which is often observed for such amorphous ZTO-based thin films as a result of film densification occurring due to more complete metal oxide formation at higher temperatures.[15]

The very similar decomposition behavior of the molecular precursors 1 and 2 inhibits the formation of any phase segregation and thus the avoidance of individual crystalline components of either zinc(II) oxide or tin(IV) oxide in the ternary ZTO film. This fact is highly conducive to the formation of the final amorphous phase within the temperature regime for thin-film processing because individual nanocrystalline zinc or tin(IV) oxide formation within the ZTO amorphous oxide matrix would deteriorate its TFT performance, as has been shown for other multinary oxides.[19,27]

A series of FET devices with a bottom-gate/bottom-contact setup were manufactured on prefabricated substrates (support/

DOI: 10.1021/acsami.7b06203
ACS Appl. Mater. Interfaces 2017, 9, 21328−21337

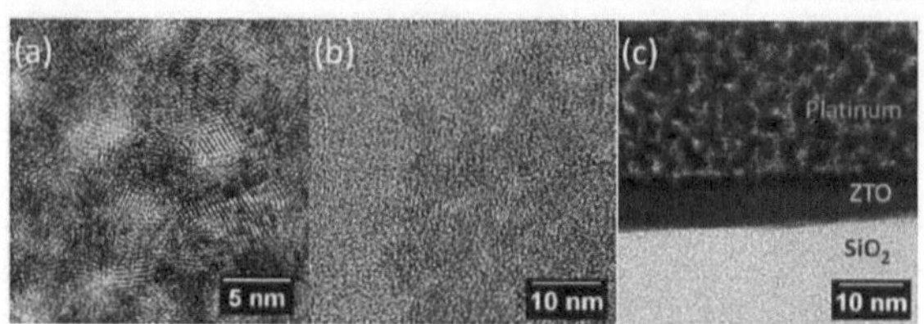

Figure 2. HRTEM images of (a) nanocrystalline SnO_2 and (b) amorphous ZTO, and (c) cross-sectional TEM (FIB) of a ZTO thin film on a Si/SiO_2 substrate. All films were finally calcined at 350 °C.

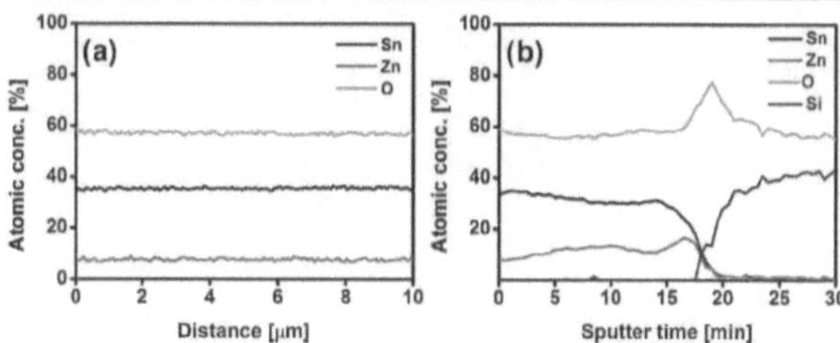

Figure 3. Auger electron spectra of (a) the surface line profile and (b) the depth profile of a ZTO thin film on a silicon substrate annealed at 350 °C.

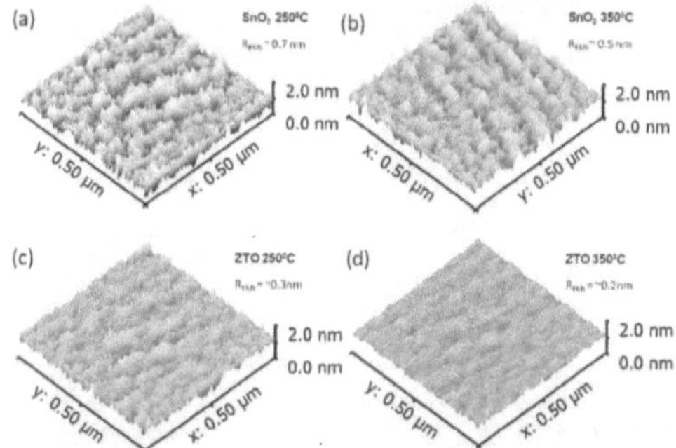

Figure 4. AFM micrographs and root-mean-square roughness (R_{RMS}) for the SnO_2 films annealed at (a) 250 °C and (b) 350 °C and ZTO films annealed at (c) 250 °C and (d) 350 °C, respectively.

gate highly n-doped silicon, dielectric silicon dioxide with interdigital gold electrodes). The ZTO films were obtained by spin coating and annealing on a hot plate at temperatures from 250 to 400 °C in air. A Sn/Zn ratio of 7:3 in the ZTO films was employed for evaluation of the FET performance.

The transfer and output characteristics of an FET device annealed at an optimum temperature of 350 °C as well as the saturation mobility (μ_{sat}), threshold voltage (V_{th}), and current on—off ratio have been determined (Figure 5 and Table 1). Active semiconductor performance was observed at all investigated temperatures (Figure S5).

Devices annealed at 300 °C and higher demonstrate a good FET performance, with charge-carrier mobilities greater than those of conventional amorphous hydrogenated silicon, which is typically ≤ 1 cm^2/V s.[1] Additionally, a systematic decrease in the threshold voltage is observed with an increase in the annealing temperature. A reduction in V_{th} generally arises from the suppression of electron traps at the interface while transitioning from low to high annealing temperatures, thereby increasing the degree of metal oxide formation. This generates the need for higher voltage sweeps to turn on the device at lower temperatures, which is in accordance with previous

DOI: 10.1021/acsami.7b06203
ACS Appl. Mater. Interfaces 2017, 9, 21328−21337

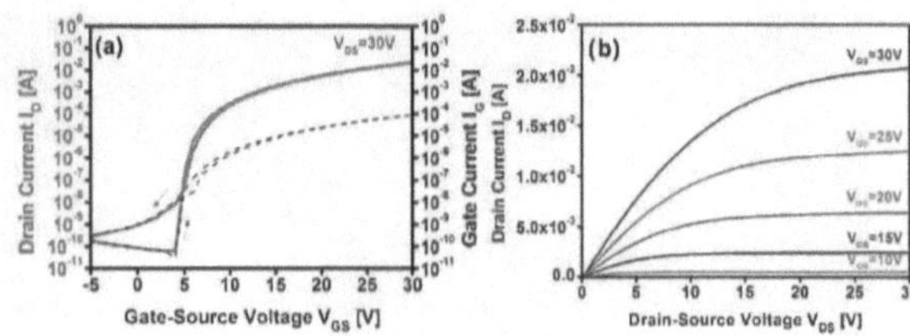

Figure 5. TFT performance characterization of the ZTO films annealed at 350 °C. (a) Transfer characteristics and (b) output characteristics of TFT devices annealed at 350 °C.

Table 1. TFT Performance Parameters for ZTO-Based Devices Annealed at Incremental Temperatures between 250 and 400 °C

temperature ($^{\circ}$C)	mobility μ_{sat} (cm^2/V s)	threshold voltage V_{th} (V)	current on/off ratio $I_{on/off}$
250	0.03	26.3	3.4×10^5
275	0.18	23.5	1.0×10^7
300	1.3	19.6	1.1×10^8
350	5.18	7.5	6.0×10^8
400	5.5	5.5	5.1×10^8

reports.[28] The dependence of the annealing temperatures and corresponding FET performance was studied to establish a suitable temperature regime below the glass-transition temperature (~450 °C) of commercially employed display-glass substrates.

Observation of active FETs at low temperatures between 250 and 300 °C, showing modest transfer characteristics, presents a future window of opportunity to obtain improved performance using further additional postprocessing treatments by calcinations under different atmospheres or using different irradiation techniques. However, within the scope of the current work, high-performance FETs were obtained at a reasonable temperature of 350 °C. These devices exhibit a beneficial reduction in the degree of hysteresis as well as a strong current saturation (Figure 5 and Table 1). Trends based on other key performance parameters of the ZTO-based FET have also been displayed (Figure S6 and Table 2). To obtain a deeper insight

Table 2. Binding Energy Trends of the Extrapolated WF and the Corresponding Valence Band Edge Maxima for the ZTO Films at Various Annealing Temperatures

annealing temperature ($^{\circ}$C)	WF (eV)	valence band maxima (VBM) edge (eV)
250	4.0	3.7
275	4.0	3.6
300	4.0	3.55
350	4.1	3.5
400	4.1	3.45

into the chemical surface environment of active ZTO thin films, XPS analysis was undertaken for the thin films annealed at different temperatures ranging between 250 and 400 °C. This should allow us to study the stepwise evolution of mixed metal oxide formation starting already from the as-deposited precursor thin film (Figure 6). Core spectra for the Sn components Sn $3d_{3/2}$ at 494.7 eV and $3d_{5/2}$ at 486.3 eV for the

350 °C films hint that Sn exists mostly in the Sn^{4+} state within the ZTO material. However an in-depth study based on Auger spectroscopy and XPS would be required to confirm the same.[29−32] In addition the O1s spectrum helps understand the nature of the different oxide species within the final material and enables an improved qualitative estimation of the electronic situation of the metal oxide.[33,34] The O1s spectra were deconvoluted with two components, arising at about 530 and 532 eV, reflecting contributions from oxidic oxygen as well as OH surface hydroxylation. Although the latter is commonly observed on most metal oxide surfaces, its prominence in comparison with the former can best be attributed either to an incomplete decomposition of the precursor or an incomplete dehydroxylation during the formation of the final ceramic.[33,35]

Additional considerations toward contributions from native point defects arising from oxygen vacancies or interstitial cationic defect sites via peak fitting did not lead to any noteworthy qualitative trends in the investigations.[34,36]

A gradual reduction in a contribution from the M−OH peak and a corresponding increase in the M−O peak were found when employing annealing temperatures between 250 and 300 °C (Figure S7). A comparatively larger change in the ratio from both contributions is observed at 350 and 400 °C (Figure 6d). Interestingly, the nature of the O1s spectra of ZTO is prominently different from those of the IZO thin films previously reported, employing the same ligand environment in the precursors used.[19] For the IZO material, within a similar temperature range, the dominant M−O metal oxide peak is observed, with a prominent M−OH peak shoulder. However, for the ZTO, this contribution, which is related to the hydroxide moiety, is rather minimal and could be observed only upon comparing the contributions from the individual oxide and hydroxide components within the final ZTO ceramic, on the basis of curve fitting of the individual components. A plausible reason for this distinct observation could be the more sluggish and thus only gradual mass loss in TGA after the initial major decomposition step has occurred for the indium precursor, which is less significant than the decomposition process observed for the herein employed tin(II) precursor. This modulation of the comparative changes in the intensity of the two major contributions serves as good insight into the evolution process of the electronic performance of metal oxide semiconductors.[19,34,37] The combined effect of determining oxygen species related to hydroxide moieties as well as the presence of both cationic and oxygen vacancies contributes toward actively changing the WF of the metal oxide semiconductor.[36]

DOI: 10.1021/acsami.7b06203
ACS Appl. Mater. Interfaces 2017, 9, 21328−21337

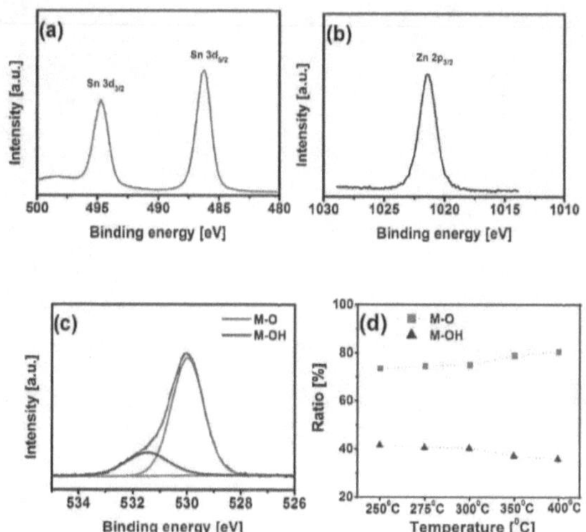

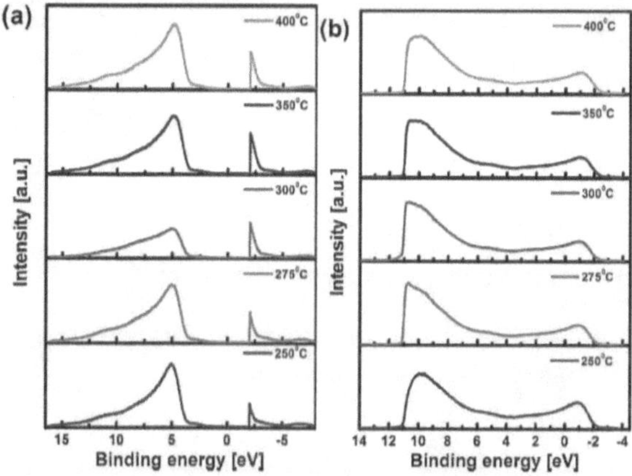

Figure 6. XPS spectra of (a) Sn 3d, (b) Zn 2p, and (c) deconvoluted O1s spectra of the deposited ZTO films annealed at 350 °C; (d) ratio of the M−O to M−OH of the ZTO thin film based on the employed calcination temperatures between 250 and 400 °C.

Figure 7. UPS spectra of ZTO thin films annealed at different temperatures. (a) WF and (b) VBM edge trends of ZTO films annealed at 250, 275, 300, 350, and 400 °C.

UPS investigations were performed on the samples also studied via XPS to obtain insight into the evolution of the VBM and the WF of the ZTO films processed in the aforementioned temperature regime (Figure 7). On the basis of the UPS measurements, samples annealed up to 300 °C exhibited a WF of ~4 eV, with a slight increase in the WF for samples annealed at 350 and 400 °C. This is in good accordance with the values reported for multinary amorphous semiconducting oxides.[38,39] The VBM edge decreases gradually with an increase in the annealing temperature, with VBM values ranging from 3.7 to

21333

DOI: 10.1021/acsami.7b06203
ACS Appl. Mater. Interfaces 2017, 9, 21328−21337

3.45 eV for samples processed at 250–400 °C, which is a good indication of successful conversion of the precursors into ZTO, which is well reflected in device performances obtained under annealing at the different temperatures studied. The VBM feature is conventionally attributed to the completely occupied 2p orbital of oxygen from the metal oxide. When analyzed together with the corresponding XPS spectra, a larger magnitude of the M–O contribution with incremental annealing temperatures has been reported to lower the VBM, resulting in a gradual decrease to lower binding energies.[1,39,40]

An increase in point defects, such as oxygen vacancies, was proposed as the most likely cause of this effect, as this would lead to an increased intrinsic charge carrier concentration, provided no additional effects should be considered.[38] This can also be correlated with the observed TFT performance, as an increased "on" current (I_{on}) and a slight reduction in the "off" current (I_{off}) was observed for TFTs processed at 350 and 400 °C, which exhibit a slightly higher WF of ~4.1 eV for both and a VBM of 3.5 and 3.45 eV, respectively. Employment of a third metal oxide or annealing performed under a dedicated oxygen-rich environment have led to improved device performance by countering the formation of undesired oxygen vacancy-related defects.[41–44]

The EPR spectra for nanocrystalline SnO_2 and ZTO powders annealed at 250 and 350 °C, respectively, give some insight into the defect situation in both materials (Figure 8). At first, the

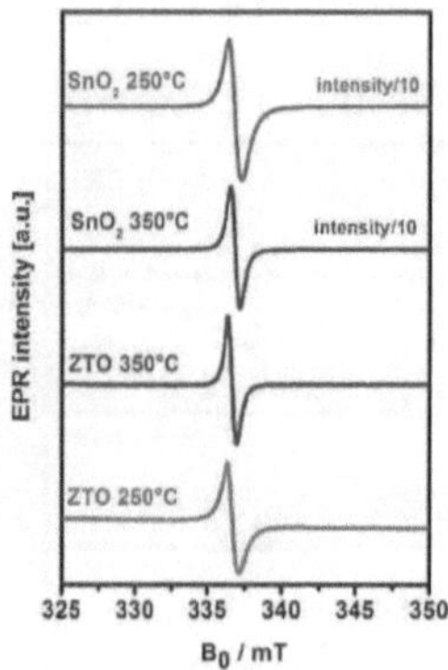

Figure 8. Room temperature X-band EPR spectra at a 325–350 mT field range for SnO_2 and ZTO, both annealed at 250 and 350 °C, respectively.

pure nanocrystalline SnO_2 powders revealed a 10 times higher intensity of the observed EPR signal than that of the amorphous ZTO powders. The EPR spectra for all samples prepared under different processing temperatures reveal an isotropic g-factor of 2.0040 for SnO_2 and 2.0032 for the ZTO samples. Slight changes in the g-factor of the nanocrystalline SnO_2 in comparison to that of the ZTO-based material might reflect differences in the spin–orbit coupling mechanism, whereas paramagnetically active electron spins arising from defects not only show interactions between different magnetic nuclei[45] (e.g., Sn and Zn) but also display a chemically different local environment. These affect the spin–orbit interactions.[46] Therefore, in the case of pure SnO_2 and mixing of Sn and Zn as in ZTO, a different environment for the defects, in particular on the surface of the materials, is established. The origin of the EPR signal can be assigned to the oxygen vacancy situation on the surface of the two materials. Thus, a different localization of oxygen vacancies changes the g-factor as well. This has already been shown previously for ZnO nanoparticles using a core–shell model, with active spin states in the bulk and on the surface of nanocrystalline ZnO particles.[23,47,48] Recently, as a proven structural model, this core–shell model for ZnO nanoparticles was developed to offer a reasonable explanation for the size effects in ZnO nanoparticles and to show that a surface shell with different types and numbers of defect sites surrounds an inner core with additional oxygen defect sites.[23,47–49] The model is also important for estimating the width of the space-charge depletion layer. According to the core–shell model, the EPR defect states with resonance at g_{iso} = 2.0040 for SnO_2 and g_{iso} = 2.0032 for the ZTO sample can be attributed to surface defect spin states. On the other hand, EPR results revealed the importance of the annealing temperature for obtaining contamination-free ZTO samples. The large sweep EPR signal (Figure S9) shows that the ZTO sample annealed at a lower temperature has a broad Gaussian line envelope, which is typical for purely carbon-related organic residuals.[50] This observation is in agreement with the lower FET performance of this particular sample, which shows this carbonaceous contamination, compared to that of the ones calcined at higher temperatures. In contrast, a similar observation of residual carbonaceous species, however, is almost negligible for nanocrystalline SnO_2 in comparison with amorphous ZTO at a lower processing temperature of 250 °C. This indicates that the contribution from the residual carbon arises most obviously from an incomplete calcination of the Zn(II) oximato precursor. This assumption is in accord with the observation from the TGA studies of the Zn(II) precursor compared with the Sn(II) precursor. The paramagnetic nature of such defect-induced carbon centers has been assigned mainly to dangling σ-bonds with an unpaired σ spin density.[51] Both samples at 350 °C exhibit a sharper and isotropic EPR line compared to that of the sample processed at 250 °C, indicating SnO_2 and ZTO materials with more uniformly distributed defect centers at higher temperatures.[52] Moreover, it was also observed that SnO_2 at 250 °C had a higher defect concentration, which was reduced at 350 °C due to thermal passivation of these defects.[53] This would indicate that defect centers, namely, oxygen vacancies of the SnO_2 sample at 350 °C, are localized in a more symmetric environment. Unlike our previous work on the In_2O_3/ZnO system,[54] herein only defects with a g-factor of 2 were observed, indicating only contributions from the surface defects and no bulk defect sites. Additionally, on the basis of our earlier EPR investigations, defects based mainly on nanoscaled ZnO were not observed in the current ZTO material system.[23] This is indicative of oxygen defects arising from the highly defective SnO_2 within the ZTO material, whereby their EPR intensity is, however, suppressed to some extent upon incorporation of a suitable amount of ZnO compared to the occurrence of the oxygen defects in the pure SnO_2 nanomaterial.

In addition to EPR spectroscopy, PL studies were performed on nanocrystalline SnO_2 and amorphous ZTO thin films calcined at 250 and 350 °C to obtain further insight into the

DOI: 10.1021/acsami.7b06203
ACS Appl. Mater. Interfaces 2017, 9, 21328–21337

contribution of optically active defects (Figure S10). The SnO_2 films calcined at the aforementioned temperatures displayed two contributions located in the UV ($\sim$340 nm) and visible ($\sim$640 nm) regions. The peak originating in the UV region of the material processed at 250 °C has been associated with quantum confinement of nanoscaled SnO_2 particles ($\sim$2–5 nm) and is absent in bulk SnO_2.[55] This observation is in good agreement with PL investigations on nanoscaled SnO_2 wires studied at a similar excitation wavelength (244 nm).[56] The contribution in the visible region can be attributed to a radiative recombination of electrons trapped at oxygen vacancy-related states of nanoscale SnO_2.[57] Interestingly, for the ZTO samples similar PL active defect contributions were observed beginning only at a processing temperature of 350 °C but not lower. This observation is in line with fact that the EPR spectra for ZTO at 250 °C show the presence of organic residues in the film, which are eliminated at a higher temperature of 350 °C. A plausible explanation for this comes from the altered decomposition process (slightly higher temperature and heating protocol, see TGA study) needed for the zinc(II) oximato precursor compared to the Sn(II) oximato precursor to undergo complete decomposition.

4. CONCLUSIONS

A new Sn(II) precursor for SnO_2 and ZTO was synthesized employing a methoxyiminopropionic acid-based ligand. In combination with the chemically analogous Zn precursor, a successful strategy to generate semiconducting ternary ZTO films and thereby functional FET devices based on amorphous ZTO thin films at reasonable temperatures between 250 and 400 °C was demonstrated. At an optimized annealing temperature of 350 °C, a high device mobility (μ_{sat}) of 5.1 cm^2/V s and a current on/off of 10^8 were obtained. The material performance derived from the employed ZTO precursor solution calcined at different temperatures was investigated with respect to its microstructure and native point defects within the final amorphous ZTO. In the present study, a clear dependency of the FET performance could be established, which is correlated with the degree of oxidation for the ZTO thin films, as determined by XPS and the dominant oxygen vacancy defect contribution, as studied by EPR and PL. These observations elucidate the prominent contributions from the defect-induced SnO_2, which play a crucial role in tuning the performance of FETs based on indium-free, semiconducting ZTO thin films. In conclusion, these investigations could provide improved insight into the semiconductor performance, on the basis of the resulting defects generated from individual precursors employed for solution processing of such multinary amorphous oxide semiconductors.

■ ASSOCIATED CONTENT

⑤ Supporting Information

The Supporting Information is available free of charge on the ACS Publications website at DOI: 10.1021/acsami.7b06203.

> Precursor studies based on TGA and IR spectroscopy, XRD of the derived SnO_2, ZTO transistor performance between 250 and 400 °C and the corresponding XPS spectra, PL spectra of SnO_2 and ZTO thin films (PDF)

■ AUTHOR INFORMATION

Corresponding Author

*E-mail: joerg.schneider@ac.chemie.tu-darmstadt.de.

ORCID ⊙

Jörg J. Schneider: 0000-0002-8153-9491

Author Contributions

The manuscript was written through contributions of all authors. All authors have given approval to the final version of the manuscript.

Funding

S.S. and J.J.S. acknowledge financial support through the DFG SPP 1569 program.

Notes

The authors declare no competing financial interest.

■ ACKNOWLEDGMENTS

TEM investigations were performed at ERC Jülich under contract ERC-TUD1. We acknowledge Dr. J. Engstler (TUDa) for performing TEM investigations, Silvio Heinschke (TUDa) for XRD analysis, and Dr. P. Atanasova (University of Stuttgart) for PL measurements. Auger measurements were carried out at Karlsruhe Nano Micro Facility (KNMF proposal number 2016-015-010549). The assistance of Tobias Weingärtner is gratefully acknowledged.

■ REFERENCES

(1) Kamiya, T.; Hosono, H. Material Characteristics and Applications of Transparent Amorphous Oxide Semiconductors. *NPG Asia Mater* **2010**, *2*, 15–22.

(2) Nomura, K.; Ohta, H.; Takagi, A.; Kamiya, T.; Hirano, M.; Hosono, H. Room-Temperature Fabrication of Transparent Flexible Thin-Film Transistors Using Amorphous Oxide Semiconductors. *Nature* **2004**, *432*, 488–492.

(3) Yabuta, H.; Sano, M.; Abe, K.; Aiba, T.; Den, T.; Kumomi, H.; Nomura, K.; Kamiya, T.; Hosono, H. High-Mobility Thin-Film Transistor with Amorphous $InGaZnO_4$ Channel Fabricated by Room Temperature Rf-Magnetron Sputtering. *Appl. Phys. Lett.* **2006**, *89*, No. 112123.

(4) Chang, Y.-J.; Lee, D.-H.; Herman, G.; Chang, C.-H. High-Performance, Spin-Coated Zinc Tin Oxide Thin-Film Transistors. *Electrochem. Solid-State Lett.* **2007**, *10*, H135–H138.

(5) Chiang, H. Q.; Wager, J. F.; Hoffman, R. L.; Jeong, J.; Keszler, D. A. High Mobility Transparent Thin-Film Transistors with Amorphous Zinc Tin Oxide Channel Layer. *Appl. Phys. Lett.* **2005**, *86*, No. 013503.

(6) Liu, L.-C.; Chen, J.-S.; Jeng, J.-S. Role of Oxygen Vacancies on the Bias Illumination Stress Stability of Solution-Processed Zinc Tin Oxide Thin Film Transistors. *Appl. Phys. Lett.* **2014**, *105*, No. 023509.

(7) Tsai, S.-P.; Chang, C.-H.; Hsu, C.-J.; Hu, C.-C.; Tsai, Y.-T.; Chou, C.-H.; Lin, H.-H.; Wu, C.-C. High-Performance Solution-Processed ZnSnO TFTs with Tunable Threshold Voltages. *ECS J. Solid State Sci. Technol.* **2015**, *4*, P176–P180.

(8) Pasquarelli, R. M.; Ginley, D. S.; O'Hayre, R. Solution Processing of Transparent Conductors: From Flask to Film. *Chem. Soc. Rev.* **2011**, *40*, 5406–5441.

(9) Jeong, S.; Jeong, Y.; Moon, J. Solution-Processed Zinc Tin Oxide Semiconductor for Thin-Film Transistors. *J. Phys. Chem. C* **2008**, *112*, 11082–11085.

(10) Lee, C.-G.; Dodabalapur, A. Solution-Processed Zinc–Tin Oxide Thin-Film Transistors with Low Interfacial Trap Density and Improved Performance. *Appl. Phys. Lett.* **2010**, *96*, No. 243501.

(11) Seo, S.-J.; Choi, C. G.; Hwang, Y. H.; Bae, B.-S. High Performance Solution-Processed Amorphous Zinc Tin Oxide Thin Film Transistor. *J. Phys. D: Appl. Phys.* **2008**, *42*, No. 035106.

(12) Chandra, R. D.; Rao, M.; Zhang, K.; Prabhakar, R. R.; Shi, C.; Zhang, J.; Mhaisalkar, S. G.; Mathews, N. Tuning Electrical Properties in Amorphous Zinc Tin Oxide Thin Films for Solution Processed Electronics. *ACS Appl. Mater. Interfaces* **2014**, *6*, 773–777.

(13) Lee, D.-H.; Chang, Y.-J.; Stickle, W.; Chang, C.-H. Functional Porous Tin Oxide Thin Films Fabricated by Inkjet Printing Process. *Electrochem. Solid-State Lett.* **2007**, *10*, K51−K54.

(14) Zhao, Y.; Dong, G.; Duan, L.; Qiao, J.; Zhang, D.; Wang, L.; Qiu, Y. Impacts of Sn Precursors on Solution-Processed Amorphous Zinc−Tin Oxide Films and Their Transistors. *RSC Adv.* **2012**, *2*, 5307−5313.

(15) Kim, Y. J.; Yang, B. S.; Oh, S.; Han, S. J.; Lee, H. W.; Heo, J.; Jeong, J. K.; Kim, H. J. Photobias Instability of High Performance Solution Processed Amorphous Zinc Tin Oxide Transistors. *ACS Appl. Mater. Interfaces* **2013**, *5*, 3255−3261.

(16) Sykora, B.; Wang, D.; Seggern, H. v. Multiple Ink-Jet Printed Zinc Tin Oxide Layers with Improved Tft Performance. *Appl. Phys. Lett.* **2016**, *109*, No. 033501.

(17) Liu, A.; Guo, Z.; Liu, G.; Zhu, C.; Zhu, H.; Shin, B.; Fortunato, E.; Martins, R.; Shan, F. Redox Chloride Elimination Reaction: Facile Solution Route for Indium-Free, Low-Voltage, and High-Performance Transistors. *Adv. Electron. Mater.* **2017**, *3*, No. 1600513.

(18) Sanctis, S.; Hoffmann, R. C.; Precht, R.; Anwand, W.; Schneider, J. J. Understanding the Temperature-Dependent Evolution of Solution Processed Metal Oxide Transistor Characteristics Based on Molecular Precursor Derived Amorphous Indium Zinc Oxide. *J. Mater. Chem. C* **2016**, *4*, 10935−10944.

(19) Hoffmann, R. C.; Kaloumenos, M.; Heinschke, S.; Erdem, E.; Jakes, P.; Eichel, R.-A.; Schneider, J. J. Molecular Precursor Derived and Solution Processed Indium−Zinc Oxide as a Semiconductor in a Field-Effect Transistor Device. Towards an Improved Understanding of Semiconductor Film Composition. *J. Mater. Chem. C* **2013**, *1*, 2577−2584.

(20) Adl, A. H.; Kar, P.; Farsinezhad, S.; Sharma, H.; Shankar, K. Effect of Sol Stabilizer on the Structure and Electronic Properties of Solution-Processed ZnO Thin Films. *RSC Adv.* **2015**, *5*, 87007−87018.

(21) Schneider, J. J.; Hoffmann, R. C.; Engstler, J.; Soffke, O.; Jaegermann, W.; Issanin, A.; Klyszcz, A. A Printed and Flexible Field-Effect Transistor Device with Nanoscale Zinc Oxide as Active Semiconductor Material. *Adv. Mater.* **2008**, *20*, 3383−3387.

(22) Schneider, J. J.; Hoffmann, R. C.; Issanin, A.; Dilfer, S. Zirconia and Hafnia Films from Single Source Molecular Precursor Compounds: Synthesis, Characterization and Insulating Properties of Potential High K-Dielectrics. *Mater. Sci. Eng., B* **2011**, *176*, 965−971.

(23) Schneider, J. J.; Hoffmann, R. C.; Engstler, J.; Dilfer, S.; Klyszcz, A.; Erdem, E.; Jakes, P.; Eichel, R. A. Zinc Oxide Derived from Single Source Precursor Chemistry under Chimie Douce Conditions: Formation Pathway, Defect Chemistry and Possible Applications in Thin Film Printing. *J. Mater. Chem.* **2009**, *19*, 1449−1457.

(24) Gerasimchuk, N.; Maher, T.; Durham, P.; Domasevitch, K. V.; Wilking, J.; Mokhir, A. Tin(IV) Cyanoximates: Synthesis, Characterization, and Cytotoxicity. *Inorg. Chem.* **2007**, *46*, 7268−7284.

(25) Stafeeva, V. S.; Mitiaev, A. S.; Abakumov, A. M.; Tsirlin, A. A.; Makarevich, A. M.; Antipov, E. V. Crystal Structure and Chemical Bonding in Tin(II) Acetate. *Polyhedron* **2007**, *26*, 5365−5369.

(26) Molloy, K. C. Precursors for the Formation of Tin(IV) Oxide and Related Materials. *J. Chem. Res.* **2008**, *2008*, 549−554.

(27) Fortunato, E.; Pimentel, A.; Gonçalves, A.; Marques, A.; Martins, R. High Mobility Amorphous/Nanocrystalline Indium Zinc Oxide Deposited at Room Temperature. *Thin Solid Films* **2006**, *502*, 104−107.

(28) Chen, H.; Rim, Y. S.; Jiang, C.; Yang, Y. Low-Impurity High-Performance Solution-Processed Metal Oxide Semiconductors Via a Facile Redox Reaction. *Chem. Mater.* **2015**, *27*, 4713−4718.

(29) Gaggiotti, G.; Galdikas, A.; Kačiulis, S.; Mattogno, G.; Šetkus, A. Surface Chemistry of Tin Oxide Based Gas Sensors. *J. Appl. Phys.* **1994**, *76*, 4467−4471.

(30) Liu, L.; An, M.; Yang, P.; Zhang, J. Superior Cycle Performance and High Reversible Capacity of SnO$_2$/Graphene Composite as an Anode Material for Lithium-Ion Batteries. *Sci. Rep.* **2015**, *5*, No. 9055.

(31) Wang, J.-J.; Lv, A.-F.; Wang, Y.-Q.; Cui, B.; Yan, H.-J.; Hu, J.-S.; Hu, W.-P.; Guo, Y.-G.; Wan, L.-J. Integrated Prototype Nanodevices Via SnO$_2$ Nanoparticles Decorated SnSe Nanosheets. *Sci. Rep.* **2013**, *3*, No. 2613.

(32) Kövér, L.; Kovács, Z.; Sanjinés, R.; Moretti, G.; Cserny, I.; Margaritondo, G.; Pálinkás, J.; Adachi, H. Electronic Structure of Tin Oxides: High-Resolution Study of XPS and Auger Spectra. *Surf. Interface Anal.* **1995**, *23*, 461−466.

(33) Abedi, N.; Heimel, G. Correlating Core-Level Shifts and Structure of Zinc-Oxide Surfaces. *Phys. Status Solidi B* **2015**, *252*, 755−764.

(34) Banger, K. K.; Yamashita, Y.; Mori, K.; Peterson, R. L.; Leedham, T.; Rickard, J.; Sirringhaus, H. Low-Temperature, High-Performance Solution-Processed Metal Oxide Thin-Film Transistors Formed by a 'Sol−Gel on Chip' Process. *Nat. Mater.* **2011**, *10*, 45−50.

(35) Kelly, L. L.; Racke, D. A.; Schulz, P.; Li, H.; Winget, P.; Kim, H.; Ndione, P.; Sigdel, A. K.; Bredas, J. L.; Berry, J. J.; Graham, S.; Monti, O. L. A. Spectroscopy and Control of near-Surface Defects in Conductive Thin Film ZnO. *J. Phys.: Condens. Matter* **2016**, *28*, No. 094007.

(36) Li, H.; Schirra, L. K.; Shim, J.; Cheun, H.; Kippelen, B.; Monti, O. L. A.; Bredas, J. L. Zinc Oxide as a Model Transparent Conducting Oxide: A Theoretical and Experimental Study of the Impact of Hydroxylation, Vacancies, Interstitials, and Extrinsic Doping on the Electronic Properties of the Polar ZnO (0002) Surface. *Chem. Mater.* **2012**, *24*, 3044−3055.

(37) Sanctis, S.; Hoffmann, R. C.; Schneider, J. J. Microwave Synthesis and Field Effect Transistor Performance of Stable Colloidal Indium-Zinc-Oxide Nanoparticles. *RSC Adv.* **2013**, *3*, 20071−20076.

(38) Kang, D.; Lim, H.; Kim, C.; Song, I.; Park, J.; Park, Y.; Chung, J. Amorphous Gallium Indium Zinc Oxide Thin Film Transistors: Sensitive to Oxygen Molecules. *Appl. Phys. Lett.* **2007**, *90*, No. 192101.

(39) Nomura, K.; Kamiya, T.; Yanagi, H.; Ikenaga, E.; Yang, K.; Kobayashi, K.; Hirano, M.; Hosono, H. Subgap States in Transparent Amorphous Oxide Semiconductor, In−Ga−Zn−O, Observed by Bulk Sensitive X-Ray Photoelectron Spectroscopy. *Appl. Phys. Lett.* **2008**, *92*, No. 202117.

(40) Socratous, J.; Banger, K. K.; Vaynzof, Y.; Sadhanala, A.; Brown, A. D.; Sepe, A.; Steiner, U.; Sirringhaus, H. Electronic Structure of Low-Temperature Solution-Processed Amorphous Metal Oxide Semiconductors for Thin-Film Transistor Applications. *Adv. Funct. Mater.* **2015**, *25*, 1873−1885.

(41) Hennek, J. W.; Smith, J.; Yan, A. M.; Kim, M. G.; Zhao, W.; Dravid, V. P.; Facchetti, A.; Marks, T. J. Oxygen "Getter" Effects on Microstructure and Carrier Transport in Low Temperature Combustion-Processed a-Inxzno (X = Ga, Sc, Y, La) Transistors. *J. Am. Chem. Soc.* **2013**, *135*, 10729−10741.

(42) Jeong, S.; Ha, Y.-G.; Moon, J.; Facchetti, A.; Marks, T. J. Role of Gallium Doping in Dramatically Lowering Amorphous-Oxide Processing Temperatures for Solution-Derived Indium Zinc Oxide Thin-Film Transistors. *Adv. Mater.* **2010**, *22*, 1346−1350.

(43) Park, S. Y.; Song, J. H.; Lee, C. K.; Son, B. G.; Lee, C. K.; Kim, H. J.; Choi, R.; Choi, Y. J.; Kim, U. K.; Hwang, C. S.; Kim, H. J.; Jeong, J. K. Improvement in Photo-Bias Stability of High-Mobility Indium Zinc Oxide Thin-Film Transistors by Oxygen High-Pressure Annealing. *IEEE Electron Device Lett.* **2013**, *34*, 894−896.

(44) Rim, Y. S.; Jeong, W. H.; Kim, D. L.; Lim, H. S.; Kim, K. M.; Kim, H. J. Simultaneous Modification of Pyrolysis and Densification for Low-Temperature Solution-Processed Flexible Oxide Thin-Film Transistors. *J. Mater. Chem.* **2012**, *22*, 12491−12497.

(45) Weil, J. A.; Bolton, J. R. *Electron Paramagnetic Resonance Elementary Theory and Practical Applications*, 2nd ed.; John Wiley: New Jersey, 2007.

(46) Pacchioni, G.; Freund, H. Electron Transfer at Oxide Surfaces. The MgO Paradigm: From Defects to Ultrathin Films. *Chem. Rev.* **2013**, *113*, 4035−4072.

(47) Kaftelen, H.; Ocakoglu, K.; Thomann, R.; Tu, S.; Weber, S.; Erdem, E. EPR and Photoluminescence Spectroscopy Studies on the Defect Structure of ZnO Nanocrystals. *Phys. Rev. B* **2012**, *86*, No. 014113.

DOI: 10.1021/acsami.7b06203
ACS Appl. Mater. Interfaces 2017, 9, 21328−21337

Research Article

(48) Parashar, S. K. S.; Murty, B. S.; Repp, S.; Weber, S.; Erdem, E. Investigation of Intrinsic Defects in Core-Shell Structured ZnO Nanocrystals. *J. Appl. Phys.* **2012**, *111*, No. 113712.

(49) Erdem, E. Microwave Power, Temperature, Atmospheric and Light Dependence of Intrinsic Defects in ZnO Nanoparticles: A Study of Electron Paramagnetic Resonance (EPR) Spectroscopy. *J. Alloys Compd.* **2014**, *605*, 34−44.

(50) Tampieri, F.; Silvestrini, S.; Ricco, R.; Maggini, M.; Barbon, A. A Comparative Electron Paramagnetic Resonance Study of Expanded Graphites and Graphene. *J. Mater. Chem. C* **2014**, *2*, 8105−8112.

(51) Boukhvalov, D. W.; Osipov, V. Y.; Shames, A. I.; Takai, K.; Hayashi, T.; Enoki, T. Charge Transfer and Weak Bonding between Molecular Oxygen and Graphene Zigzag Edges at Low Temperatures. *Carbon* **2016**, *107*, 800−810.

(52) Repp, S.; Weber, S.; Erdem, E. Defect Evolution of Nonstoichiometric ZnO Quantum Dots. *J. Phys. Chem. C* **2016**, *120*, 25124−25130.

(53) Kamble, V. B.; Bhat, S. V.; Umarji, A. M. Investigating Thermal Stability of Structural Defects and Its Effect on D0 Ferromagnetism in Undoped SnO_2. *J. Appl. Phys.* **2013**, *113*, No. 244307.

(54) Hoffmann, R. C.; Kaloumenos, M.; Heinschke, S.; Erdem, E.; Jakes, P.; Eichel, R.-A.; Schneider, J. J. Molecular Precursor Derived and Solution Processed Indium-Zinc Oxide as a Semiconductor in a Field-Effect Transistor Device. Towards an Improved Understanding of Semiconductor Film Composition. *J. Mater. Chem. C* **2013**, *1*, 2577−2584.

(55) Lee, E. J. H.; Ribeiro, C.; Giraldi, T. R.; Longo, E.; Leite, E. R.; Varela, J. A. Photoluminescence in Quantum-Confined SnO_2 Nanocrystals: Evidence of Free Exciton Decay. *Appl. Phys. Lett.* **2004**, *84*, 1745−1747.

(56) Mazeina, L.; Picard, Y. N.; Caldwell, J. D.; Glaser, E. R.; Prokes, S. M. Growth and Photoluminescence Properties of Vertically Aligned SnO_2 Nanowires. *J. Cryst. Growth* **2009**, *311*, 3158−3162.

(57) Shewale, P. S.; Ung Sim, K.; Kim, Y.-b.; Kim, J. H.; Moholkar, A. V.; Uplane, M. D. Structural and Photoluminescence Characterization of SnO_2: F Thin Films Deposited by Advanced Spray Pyrolysis Technique at Low Substrate Temperature. *J. Lumin.* **2013**, *139*, 113−118.

DOI: 10.1021/acsami.7b06203
ACS Appl. Mater. Interfaces 2017, 9, 21328−21337

6.5 Aqueous Solution Processing of Combustible Precursor Compounds Into Amorphous Indium Gallium Zinc Oxide (IGZO) Semiconductors for Thin Film Transistor Applications

Aqueous combustion synthesis has gained much attention as an environmentally friendly and effective strategy for low-temperature solution-processing of oxide semiconductors. The combination of specific amounts of "oxidizer" (e.g. metal nitrates) and "fuel" (e.g. urea or acetylacetone) results in exothermic combustion reactions, enabling the complete conversion of the precursor at lower processing temperatures. Unfortunately, the ageing of the chemical species in solution over time render the reproducibility complicated and leading to a deficient understanding of the decomposition process. A possible strategy to overcome this issue is to design well-defined precursor compounds with the fuel and oxidizer components available within one precursor compound. This eliminates the need of any additives or stabilizers prior to the direct solution processing of the precursor thin films for the formation of oxide semiconductors at low processing temperatures.

In this publication we report on the synthesis and full structural characterization of well-defined urea nitrate coordination compounds of indium (III), gallium (III) and zinc (II) for the formation of semiconducting IGZO thin films and its application in transistor devices. The single-source molecular precursors were extensively characterized by spectroscopic techniques as well as by single-crystal structure analysis. The thermal decomposition behavior of the respective precursors was investigated by a TG-MS/IR. In all cases the precursors thermally decompose in multiple steps involving complex redox-reactions, whereby the *in situ* formation of nitrogen containing molecular species can be detected. The exothermic decomposition of all three precursor compounds was confirmed by DSC analysis displaying multiple sharp exothermic peaks between 200 and 400 °C as a result of the combustion synthesis emerging from the urea-nitrate (fuel-oxidizer) reaction. As a result, a controlled thermal transformation of a defined mixture of the indium, gallium and zinc urea nitrate coordination compounds into quaternary amorphous IGZO thin films could be accomplished. In addition, XPS investigations reveal a clean decomposition for IGZO films processed between 300 and 350 °C with no significant amounts of residual hydrocarbon or nitrogen species in the final films.

Thin-film transistors (TFTs) were fabricated from a defined compositional mixture of the single-source molecular precursors (In:Ga:Zn, 7:1:1.5). The fabricated TFTs showed an active transistor performance even at processing temperatures as low as 200 °C. TFT devices processed at 300 °C and 350 °C exhibit good electrical performances with saturation mobilities μ_{sat} of 1.7 and 3.1 cm^2 V^{-1} s^{-1} and a threshold voltage V_{th} of 8.6 and 4.3 V, respectively as well as current on-off ratios $I_{on/off} > 10^7$ in both cases.

DOI: 10.1002/asia.201801371

CHEMISTRY
AN ASIAN JOURNAL
Full Paper

Thin Film Transistors

Aqueous Solution Processing of Combustible Precursor Compounds into Amorphous Indium Gallium Zinc Oxide (IGZO) Semiconductors for Thin Film Transistor Applications

Shawn Sanctis,[a] Rudolf C. Hoffmann,[a] Nico Koslowski,[a] Sabine Foro[++,][c] Michael Bruns[+,][b] and Jörg J. Schneider*[a]

Abstract: Combustion synthesis of semiconducting amorphous indium gallium zinc oxide IGZO (In:Ga:Zn, 7:1:1.5) thin films was carried out using urea nitrate precursor compounds of indium(III), gallium(III) and zinc(II). This approach provides further understanding towards the oxide formation process under a moderate temperature regime by employment of well-defined coordination compounds. All precursor compounds were fully characterized by spectroscopic techniques as well as by single crystal structure analysis. Their intrinsic thermal decomposition was studied by a combination of differential scanning calorimetry (DSC) and thermogravimetry coupled with mass spectrometry and infrared spectroscopy (TG-MS/IR). For all precursors a multistep decomposition involving a complex redox-reaction pathway under in situ formation of nitrogen containing molecular species was observed. Controlled thermal conversion of a mixture of the indium, gallium and zinc urea nitrate complexes into ternary amorphous IGZO films could thus be achieved. Thin film transistors (TFTs) were fabricated from a defined compositional mixture of the molecular precursors. The TFT devices exhibited decent charge carrier mobilities of 0.4 and 3.1 cm^2/(Vs) after annealing of the deposited films at temperatures as low as 250 and 350 °C, respectively. This approach represents a significant step further towards a low temperature solution processing of semiconducting thin films.

Introduction

Amorphous oxide semiconductors (AOS) are among the most rapidly developing classes of materials used for electronic and optoelectronic applications.[1] The intrinsic properties of these wide band gap semiconductors like their almost complete optical transparency in the visible region and their high field effect mobilities recommend them for use in the field of next generation display technology.[2]

Currently the key players are multinary metal oxides, namely indium zinc oxide (IZO), indium gallium zinc oxide (IGZO) or zinc tin oxide (ZTO) which are mainly deposited by means of vacuum-based techniques.[3] Although the solution processing of AOS is possible and comes with the promise of several advantages such as low costs, large area and roll-to-roll deposition[4] as well as direct photo-patterning,[5] these methods did not win sufficient recognition for technological application so far due to the still unresolved inherent problems concerned with their use. Although molecular precursor compounds which start to decompose conveniently at low temperatures of ≈200 °C, unfortunately still contain traces of organic residues from the ligand sphere around the metal center, which remain up to significant higher temperatures in the thin oxide films that is close to 400–500 °C. Typically these carbonic residues severely corrupts the electronic performance.[6] The complete removal of the organic remains thus requires the sufficient supply of atmospheric oxygen and longer annealing times or elevated processing temperatures.[7] Several strategies exist to cope with this drawback. One approach uses the systematic introduction of reactive ligand groups, such as nitro or nitroso functionalities, into the ligand framework of the precursor. This facilitates the start of the decomposition via thermal decay.[8] Another attempt utilizes a so-called "combustion processing" which allows complete conversion of the precursor at lower temperatures by providing specific amounts of "oxidizer" (e.g. metal nitrates) and a "fuel" (e.g. urea or acetylacetone).[9] This

[a] S. Sanctis, Dr. R. C. Hoffmann, N. Koslowski, Prof. J. J. Schneider
Department of Chemistry, Eduard-Zintl Institut für Anorganische und Physikalische Chemie
Technische Universität Darmstadt
Alarich-Weiss-Straße 12, 64287 Darmstadt (Germany)
E-mail: joerg.schneider@ac.chemie.tu-darmstadt.de

[b] Dr. M. Bruns[+]
Institute for Applied Materials-Karlsruhe Nano Micro Facility
Karlsruhe Institute of Technology (KIT)
Hermann-von-Helmholtz-Platz 1, D-76344, Eggenstein-Leopoldshafen (Germany)

[c] S. Foro[++]
Department of Material Science
Technische Universität Darmstadt
Alarich-Weiss-Straße 8, 64287 Darmstadt (Germany)

[++] X-ray crystal structure determination.

[+] XPS investigation.

Supporting information and the ORCID identification number(s) for the author(s) of this article can be found under:
https://doi.org/10.1002/asia.201801371.

two-component concept has demonstrated remarkable results, when transferred to the synthesis of AOS.[3a–c] Nevertheless, further insight into the decomposition process occurring during the combustion synthesis leading to metal oxides is needed. The aim of the current work is to introduce a new approach for studying the interplay between the metal nitrate and the urea ligands. This is achieved by introducing the concept of defined molecular precursors. Recent investigations of aqueous solutions of metal nitrate employing urea as an additive has shown the need for appropriate ageing of the solution over several days. The so far described ageing process ensures the in situ complexation of the urea to the metal cation, to obtain the desired combustion effect. Nevertheless this process calls for a controlled solution processing and is vulnerable to problems in processing[10] Herein we have chosen a single source precursor approach in which we combine the urea ligands (the fuel) and the nitrate ions (the oxidizer) in one specific compound. This enables us to elucidate their role in the formation of the desired metal oxides more precisely in a controlled manner.[11] During the thermal decomposition and the subsequent comproportionation, reaction between the nitrogen containing constituents (i.e. nitrate anion and urea) occurs. Hereby, a complex mixture of compounds with nitrogen in various oxidation states can be expected.[12]

In the current work we demonstrate the synthesis and characterization of well-defined urea nitrate coordination compounds of indium(III) (**1**), gallium(III) (**2**) and zinc(II) (**3**) and their use as single source precursors to access the formation of amorphous IGZO thin films. The molecular structures of compounds (**1**)–(**3**) were determined by single crystal X-ray diffraction. A detailed analysis of the thermal conversion of the molecular precursors (**1**)–(**3**) to yield the metal oxides was carried out using TG and DSC techniques as well as hyphenated techniques for identifying gaseous by-products by IR and MS.

IGZO based thin films were obtained by spin-coating of the molecular precursors using water as an eco-friendly solvent resulting in the formation of amorphous, smooth homogeneous semiconducting oxide thin films at temperatures between 200 °C to 350 °C which exhibit an excellent TFT transistor behaviour. X-ray photoelectron spectroscopy elucidated the improvement in TFT performance is based largely on the improved metal oxide formation as well as passivation of defects within the local oxygen environment of the metal oxide microstructure.

Results and Discussion

Synthesis and thermal analysis

The precursor complexes [M(urea)$_6$][NO$_3$]$_3$ with M = In (**1**) and Ga (**2**) as well as [Zn(urea)$_4$(H$_2$O)$_2$][NO$_3$]$_2$ (**3**) were synthesized by reaction of the metal nitrate hydrates with stoichiometric amounts of urea in ethanol for (**1**) and (**2**) and n-butanol (**3**) as solvent as described in Figure 1. Subsequent recrystallization can be performed from methanol, by a modified procedure.[13] Compounds (**1**) and (**2**) were isotypic and crystallize in the space group C2/C, whereas crystals of (**3**) exhibited the space

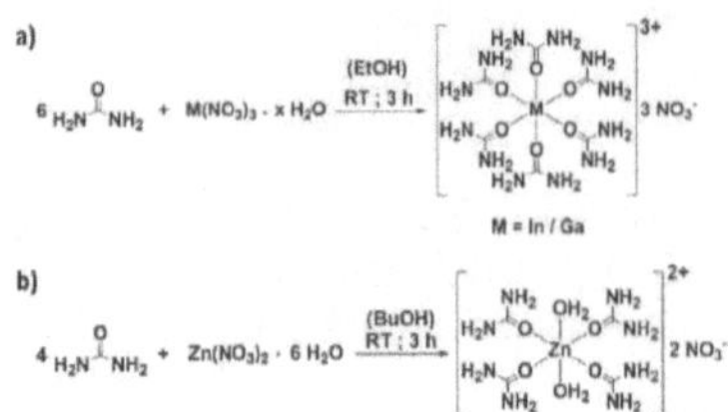

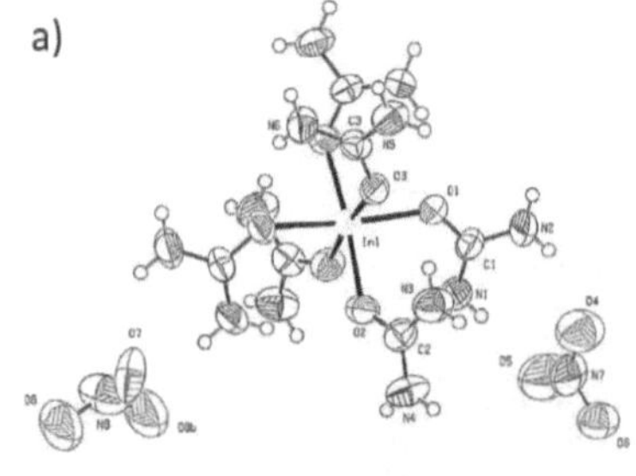

Figure 1. Schematic representation for the synthesis of the metal urea complexes of indium (**1**), gallium (**2**) and zinc (**3**) respectively.

group $P2_1/n$ (Figure 2). Hereby, mononuclear complexes with octahedral coordination of the central metal atoms were observed in all cases. The neutral urea molecule serves as monodentate ligand and coordinates to the central metal atom by its oxygen atom. For (**1**) and (**2**) the trivalent In and Ga metal

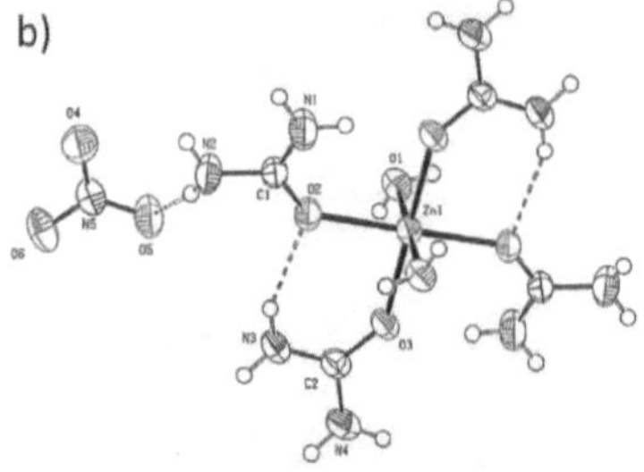

Figure 2. ORTEP plot of the metal urea complexes of a) [In(urea)$_6$][NO$_3$]$_3$ (**1**) and b) [Zn(urea)$_4$(H$_2$O)$_2$][NO$_3$]$_2$ (**3**) drawn at a 50% probability level. (Both crystallographically independent nitrate anions in (**1**) exhibited structural disorder, which in each case was described using two split position. For clarity only one orientation is thus shown. The third nitrate anion is related to the former by crystal symmetry and therefore not shown either.

atoms were homoleptically coordinated by six urea molecules, whereas for (**3**) the coordination sphere consisted of four urea and two water molecules completing the octahedral coordination sphere around the zinc center in (**3**). Weak hydrogen bonds between the urea ligand and the nitrate anions were observed for (**3**), which probably prevents rotational disorder of the nitrate anion in the case. Further details of the complete analytical data of (**1**) can be found in Tables S1 and S2.[37] Such hexakis(urea) compounds with trivalent metal ions are for example, known for chromium,[14] or manganese.[15] Although, the molecular structure of the divalent zinc compound (**3**) has been previously reported,[37] a straight forward procedure to isolate the final product was not established so far.[16] This warrants a report herein. Although homoleptic urea complexes of the composition $M(urea)_6^{n+}$ as observed for (**1**) and (**2**) are quite common[16,17] and under non aqueous reaction conditions a homoleptic zinc coordination $[Zn(urea)_6][NO_3]_2$ was reported,[18] a general composition $M(urea)_4(H_2O)_2^{n+}$ as found for (**3**) is much less common so far.[19] For the Ga compound (**2**) the obtained crystal structure shows a similar co-ordination of the urea ligand to the metal centre as observed for (**1**) (Figure S1). However, the crystal structure refinement was poor and thus unsuitable for a detailed refinement. The connectivity of the atoms for (**2**) could however be established unequivocally.

The influence of the counter anion for such metal urea complexes also plays a crucial role in the nature of their thermal decomposition, whereby the urea complexes with a nitrate anion undergo an enhanced thermal conversion to the ceramic oxide at much lower temperatures due the "fuel/oxidizer" reaction between the urea and the nitrogen containing species (NO_2, HONO) compared to their chloride counterparts. The need of elevated temperatures for urea complexes of metal chlorides is largely due to an endothermic reaction leading to the formation of stable intermediates which undergo further decomposition only at elevated temperatures ($>600^{\circ}C$) accompanied by the elimination of hydrochloric acid.[11–12] Figure 3 shows TG mass loss (a) and the corresponding DSC studies (b-

d) of the urea-nitrate complexes (**1**), (**2**) and (**3**) in oxygen. Although, complexes (**1**), (**2**) and (**3**) possess identical urea/nitrate ratios of 6:3 for (**1**), (**2**) and 4:2 for (**3**), they exhibit very different mass loss curves (Figure 3 a, Figure S2). All three precursors depict a prominent starting decomposition step at about 175 °C with a slight initial mass loss for (**3**) which corresponds to the loss of the two co-ordinated water molecules. Initial mass loss at lower temperatures (<100 °C) resulting from the release of the coordinated water molecules in form of vapor is well known for zinc acetates[20] and zinc acetylacetonates[21] and is in good agreement with our observations.

DSC analysis (Figure 3 b–d) confirms the exothermic decomposition of all three precursors with a significantly sharp exothermic peak between 100 °C and 200 °C followed by a less intense exothermic peak between 250 °C. Minor signals of endothermic events in the case of complexes (**1**) and (**3**) can be associated with removal of residual carbonaceous species. The thermal decomposition of (**1**)–(**3**) was carried out in oxygen to get an insight in the decomposition by-products of the precursors. The decomposition behavior of the indium and zinc precursors proceeded as a two-step decomposition where significant mass loss occurs in the first step itself. The residual mass of all the precursors are in good agreement with the expected ceramic yield from the decomposition of the precursors in oxygen ((**1**): $CY_{calc.}$ 21.3%, $CY_{meas.}$ 21.8%; (**2**): $CY_{calc.}$ 15.2%, $CY_{meas.}$ 15.3%, (**3**): $CY_{calc.}$ 17.5% and $CY_{meas.}$ 18.4%. The differential scanning calorimetry analysis of these precursors also indicate a sharp exothermic decomposition which can be associated with the expected redox reactions between the co-ordinated urea and the nitrate anion.[22]

Thermal analysis of the evolved gaseous by-products from the combustion-based decomposition were analyzed by means of TG-MS-FTIR (Figures 4 and 5).

The majority of the observed gases were largely identical where the resultant gases during the combustion process could be assigned to ammonia (m/z^+ 17), water (m/z^+ 18), carbon monoxide or nitrogen (m/z^+ 28), nitric oxide (m/z^+ 30), isocyanic acid (m/z^+ 43) and nitrogen dioxide (m/z^+ 46). The

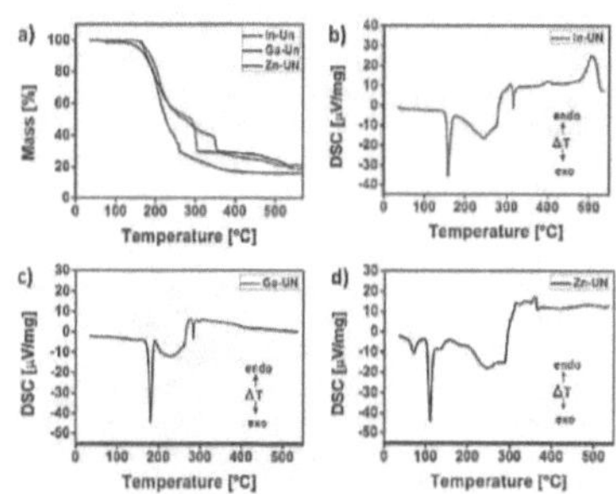

Figure 3. (a) TG mass loss and (b–d) corresponding DSC studies of the urea-nitrate complexes (**1**), (**2**) and (**3**) in oxygen.

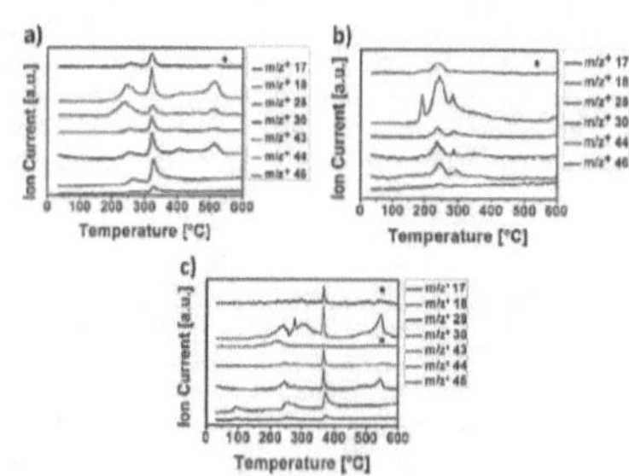

Figure 4. MS intensities of a) Indium precursor b) Gallium precursor and c) Zinc precursor for $m/z+$ peaks corresponding to the TG curves in Figure 3 respectively. (trace marked with * is enhanced by a factor of 10 for clarity).

Chem. Asian J. **2018**, *13*, 3912–3919   www.chemasianj.org   3914

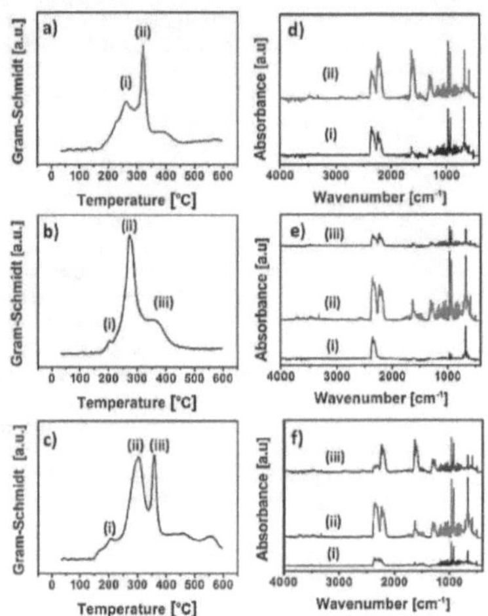

Figure 5. Gram-Schmidt intensities (a–c) of the indium, gallium and the zinc urea-nitrate precursors (1)–(3) and the corresponding IR signal intesities (d–f) according to the Gram-Schmidt the precursors, respectively.

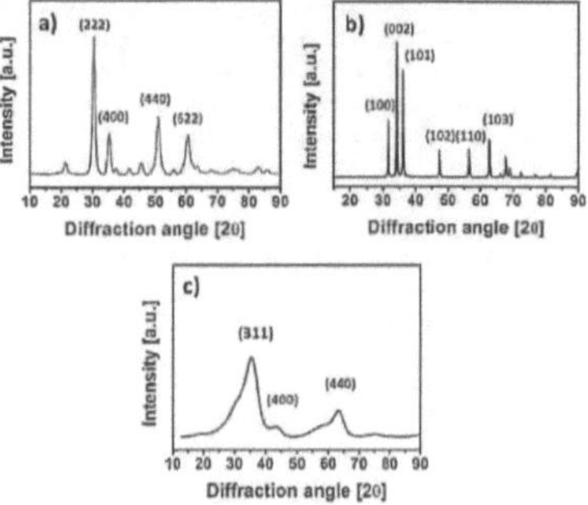

Figure 6. XRD analysis of the metal oxides powders obtained from the thermal combustion of aqueous urea nitrate precursors (1)–(3) and final calcination at 350 °C for 2 hours, respectively.

signal corresponding to (m/z^+ 44) can be correlated with the evolution of carbon dioxide or nitrous oxide. The IR signals corresponding to the different signal intensities of the Gram-Schmidt signal were recorded to identify the various gases being released as a result of the precursor combustion. For all the three precursors, a clear evolution of ammonia ($NH_3 = 968$ cm^{-1}) isocyanic acid ($HNCO = 2283$ and cm^{-1}), carbon-dioxide ($CO_2 = 2332$ and 2362 cm^{-1}), nitric acid ($HNO_3 = 1713$ and 1325 cm^{-1}) and nitrous oxide ($N_2O = 2241$ cm^{-1}) were detected at various decomposition stages ranging between 200 °C and 350 °C. This is in good agreement with observed thermal decomposition products of urea complexes with metal nitrates.[23]

The deduction of a detailed mechanism for the thermal decay of (1)–(3) remains difficult though. Clearly consecutive processes occur in all cases, but the determinable gaseous reaction products from MS or FTIR do not vary significantly to securely trace down a detailed decomposition pathway.

In order to elucidate the nature of the resultant ceramic oxide after decomposition of precursors (1)–(3) they were dissolved each in water, subsequently spin-coated on quartz substrates and calcined followed by isolation the calcined AOS material. In addition, thin films of the individual precursors calcined each at 350 °C allow the formation of the individual oxides of indium, zinc and gallium displaying a crystalline nature corresponding to the bixbyite (for In_2O_3) and wurtzite (ZnO) structures as shown in Figure 6. Interestingly, gallium oxide exhibits a weak crystalline phase with broad reflexes which corresponds to the "δ-Ga_2O_3" phase.[24] Annealing pre-

cursor (2) at even higher temperature of 800 °C revealed a clear formation of phase pure monoclinic β-Ga_2O_3 with no detection of any other mixed secondary phase Figure S3. It is known that gallium oxide displays a broad phase distribution depending on the type of precursor used and the thermal calcination history under which this material is obtained. This allows to form different crystalline phases or even mixtures therefrom. Indepth neutron diffraction studies have previously indicated that the "δ-Ga_2O_3" is actually a mixture of hexagonal ε-Ga_2O_3 and β-Ga_2O_3.[25]

AOS thin film characterization

For multinary oxide semiconductor films, it is essential to have the final ceramic film in the amorphous state to avoid any charge transport barriers arising from grain boundaries from the polycrystalline nanoparticles dispersed within the amorphous matrix. In order to gain an insight into the fabricated IGZO film, precursor solution was generated by mixing the individual precursors for (1), (2) and (3) and spin coated on Si/SiO$_2$ substrates. The aqueous IGZO precursor solutions were spin-coated to generate well-coated thin-films and annealed between 200 °C and 350 °C for 2 hours, respectively. Three iterations of the film-coating procedure were repeated in order to obtain the desired film thickness of the IGZO films for which their transistor performance was assessed later. Attempts to directly spin coat aqueous IGZO solutions from pure metal nitrates analogous to the urea nitrate precursor compounds resulted in inhomogeneous film coatings with poor adhesion to the SiO$_2$ dielectric and could not be analyzed for further investigations. Similar inconsistencies have been previously reported, which were circumvented by the formation of electrochemically generated aqueous metal-hydroxo nanoclusters.[26] The homogeneity of the coated films from the urea nitrate precursors on a Si/SiO$_2$ substrate as well as their microstructure was investigated via high-resolution transmission electron microscopy (HR-TEM) and atomic force microscopy (AFM). The TEM investigations (Figure 7 a,b) revealed that films annealed at both temperatures are amorphous in nature with the absence of any partial film crystallinity amorphous matrix.[27]

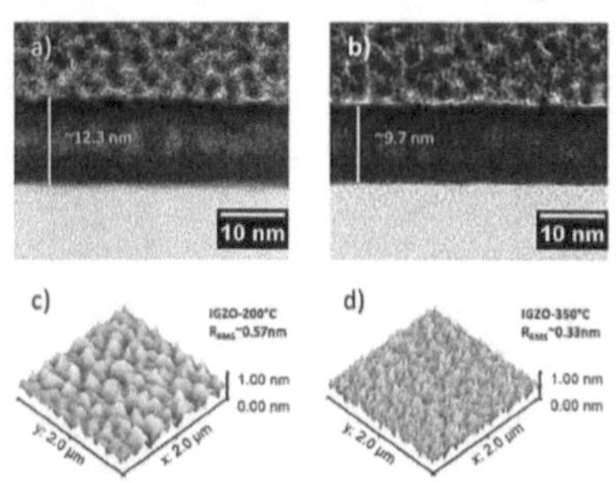

Figure 7. Cross-sectional HR-TEM images with measured film thickness and AFM micrographs and root-mean-square roughness (R_{RMS}) for the IGZO films annealed at (a, c) 200 °C and (b, d) 350 °C respectively.

This serves as a good indication that film annealing at elevated temperatures actively promotes film densification, thereby improving its contact/adhesion with the underlying dielectric layer (SiO$_2$), which plays a crucial role in eliminating void formation which are usually a consequence of residual organic elimination, and evolution of gaseous by-products.[28] The resultant densification of the film is known to improve the electrical performance of TFT devices by having a good adhesion and as well as passivation of interfacial defects.[29]

AFM of the films annealed between 200 °C and 350 °C were analyzed to understand the morphology of the calcined IGZO films (Figure 7 c,d). Films annealed at the lower temperature display a more textured surface indicating a relatively rougher film in comparison with films calcined at 350 °C which show a smoother morphology. This is well corroborated with the results obtained from the HR-TEM investigations.

Thin film transistor performance

In order to study the influence of the IGZO thin films, a series of TFT were produced in a bottom gate bottom contact (BGBC) configuration. Device structures with a reasonably large source-drain electrode geometry with a width (W) to length (L) ratio (W/L = 500) were utilized to in order to avoid any overestimation of device mobility. It must be noted that all devices were fabricated with a standard SiO$_2$. Higher device performances can be typically achieved using high-k dielectrics oxides.[30] Thin films of IGZO were generated in the source drain area of the TFT by spin-coating of an aqueous precursor solution containing (1), (2), and (3) followed by drying and combustion synthesis and finally annealing between 200 °C and 350 °C for two hours. The essential performance metrics of the transis-

tors, such as the saturation mobility ($\mu_{sat.}$), threshold voltage (V_{th}) and the current on-off ratio were measured and are shown below Figure 8.

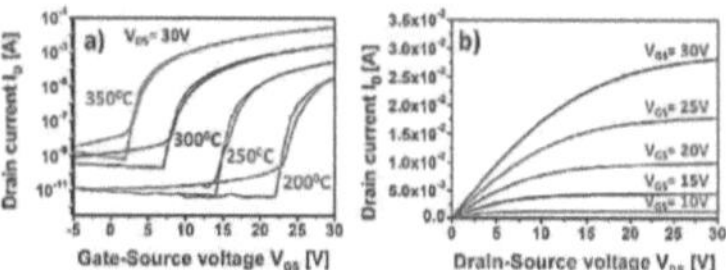

Figure 8. TFT performance characterization based on IGZO thin films annealed at various temperatures. a) Comparative transfer characteristics of IGZO TFTs annealed between 200 °C and 350 °C. b) Output characteristics of the device annealed at 350 °C.

Active transistor performance was observed at all the investigated temperatures and the corresponding transistor performance parameters have been summarized in Table 1.

Table 1. TFT performance parameters of IGZO based devices annealed at incremental temperatures between 200 and 350 °C.

Annealing Temp.	Mobility(μ) cm^2/[Vs]	V_{th} [V]	$I_{on/off}$
200 °C	0.09	23	2.3×10^5
250 °C	0.4	17.4	6.4×10^6
300 °C	1.7	8.6	4.1×10^7
350 °C	3.1	4.3	8.4×10^7

TFTs annealed at 300 °C and 350 °C exhibit a good device performance with charge-carrier mobilities μ_{sat} of 1.7 cm^2/(V·s) and 3.1 cm^2/(V·s), respectively as well as current on-off ratios of $>10^7$ in both cases. It is motivating to observe that devices annealed at temperatures at 200 °C and 250 °C still show active semiconductor behavior although the performance in not on par with the that of the devices annealed at higher temperatures, with relatively higher (Vth) and lower on current in the transfer characteristics, accompanied by lower gate- voltage modulation, with a lower potential for device applications. However, devices annealed at higher temperatures show a good mobility > 1 cm^2/(V·s). Interestingly, similar trends in increased mobility and negative shifts of the threshold voltages at much higher annealing temperatures, were previously reported for solution processed IGZO films from nitrate salts. The observed improvement in TFT performance was attributed to the formation of oxygen vacancy related point defects at elevated temperatures which enhances the generation of addition electron charge carriers. This in turn facilitates the improvement device mobility as well as threshold voltage.[31]

X-ray photoelectron spectroscopy analysis for the IGZO thin films were carried out in order to probe into the change in the nature of the chemical species involved. The oxygen O1s spectra serves a good indicator of the enhanced formation of the ceramic oxide, thereby facilitating a good understanding of the microstructure -property relation of the metal oxide. The deconvoluted oxygen O1s spectra of the IGZO films annealed at different temperatures are shown in Figure 9. Curve fitting of the oxygen peak was carried out using contributions with positions centered at ≈ 530.2 eV, ≈ 531.8 eV and ≈ 532.8 eV which are indicative of the different oxygen environments. Quantitative contributions from the different oxygen environments with respect to their atomic percentages can be found in Table S3.

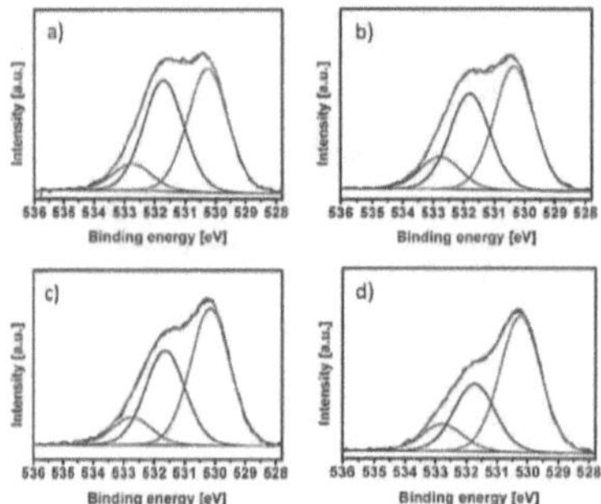

Figure 9. Deconvoluted O1s XPS spectra (a–d) of the IGZO films annealed at 200 °C, 250 °C, 300 °C, 350 °C respectively.

The peak position centered ≈ 530.2 eV indicates metal-oxygen bonds (M-O) within the IGZO bulk representing a complete coordination environment by O^{2-} ions. The peak at ≈ 531.8 eV can be attributed to oxide O^{2-} associated with hydroxide species which are coordinated to the metal atoms (M-OH) as well as possible contributions from surface adsorbates.[32] The third contribution to the O1s peak is positioned at an even higher binding energy of ≈ 532.8 eV and might well originate from oxygen species from the underlying SiO_2 dielectric layer which is known to create an intermixed phase close to the boundary where Si^{4+} diffuses in the Si/IGZO interface.[29] This corroborates well with the ultra-thin films (< 20 nm) as well as its dense morphology (see Figure 7).[29–31,33] Although, such contributions are also known to be due to oxygenated carbon species arising from a topmost contamination layer (e.g. carbonyl, carbonate as well as C-OH groups),[3a,c,32,34] it seems rather less likely for the investigated samples. The shoulder at higher binding energies in the O1s peak was almost similar across all samples with no significant decrease at higher annealing temperatures. Such a thermal treatment should affect surface contaminations due to the ease of their thermal removal. Subsequently, it did not affect the qualitative contributions from the other binding energies (see Suppl. Table S3). A comparative increase in the ratio of the M–O bonds and the simultaneous decrease in the M-OH bonds is clearly observed, which further confirms a thermally assisted densification with a possible interdiffusion of Si^{4+} into the

IGZO phase. The charge carrier transport in such amorphous oxide semiconductors is directed through the improved orbital overlap of large vacant s-orbitals at the conduction band minimum, which originates from the co-ordinated M–O bonds in such ionic oxides.[35] This is in good agreement with increase in TFT performance, especially in terms of improved charge carrier mobility of the devices annealed at higher temperatures which generates an improved M-O-M network, accompanied by better percolation pathways for charge carriers in relatively dense films with reduced porosity.[28b, 29] Recent investigations consisting UV irradiation of metal nitrate precursors have been explored to produce high performance amorphous semiconducting oxides from a solution process.[36] Such an approach could be attempted in the future, since the urea complexes of metal nitrates are also compounds with the inherent possibility of initiating the combustion reaction via UV processing.

Conclusions

Eco-friendly water-soluble metal urea complexes of indium, gallium and zinc precursors of composition In/Ga(urea)$_6$(NO$_3$)$_2$ and Zn(urea)$_4$(H$_2$O)$_2$(NO$_3$)$_2$ were successfully employed in the generation of active semiconductor IGZO thin films via a solution-processing route. The precursors undergo a desirable exothermic decomposition (combustion) when annealed at relatively low temperatures between 200 and 350 °C, with volatile gaseous by-products affording the formation of clean ceramic oxides which enable the formation of amorphous IGZO thin films. Integration of such solution-processed thin-films enables the formation of active semiconductor materials. At an optimum temperature of 350 °C, TFT performance with the device mobility (μ_{sat}) of 3.1 cm^2/Vs and a current on/off ratio of 10^7 was achieved. The formation of the thin-film oxides from the urea nitrate complexes was investigated with respect to their microstructure-property relationship and the change in the oxygen environment at different annealing temperatures by XPS. HR-TEM and AFM investigations show a visible dependency of the film density and surface morphology, wherein dense and smooth films could be manufactured at reasonable calcination temperatures. This was further confirmed by an improved degree of ceramisation exhibiting a reduced presence of hydroxide species at elevated calcination temperatures. This observation is well corroborated with the observed TFT performance in the IGZO films derived from the urea nitrate complexes.

Experimental Section

Precursor synthesis and analysis

Hexakis(urea)indium(III) nitrate (1): Indium nitrate hydrate (3 g, 10 mmol) was dissolved in 100 mL of anhydrous ethanol. An excess of urea (5.40 g, 90 mmol) was added thereafter into the solution. The solution was stirred for 3 hours and the formation of a visible white precipitated product is observed. The product was washed in ethanol to ensure the removal of any by-products and then dried in vacuo (7.45 g, 84%). The compound is soluble in water and moderately soluble in methanol. Crystals suitable for X-

ray crystal structure determination were obtained by creating a saturated solution in methanol and layering with diethylether for several days at room temperature. Anal. calcd (%) for $C_6H_{24}InN_{15}O_{15}$ (661.17 g mol^{-1}) C 10.90 H 3.66 N 31.78; found C 10.64 H 3.76 N 31.57. ^{1}H NMR ([D$_6$] dimethyl sulfoxide, 25 °C) $\delta = 5.49$ ppm (br, -NH$_2$). ^{13}C{^{1}H} NMR ([D$_6$] dimethyl sulfoxide, 25 °C): 159.67 (C=O). IR (crystals) $\tilde{\nu} = 3431$ (s, $\nu_{NH,out\ of\ phase}$), 3343 (s, $\nu_{NH\ in\ phase}$), 3232 (s, $\nu_{NH,ip}$), 1630 (s, ν_{CO}), 1557 (s, δ_{NH}), 1504 (s, δ_{NH}), 1329 (s, ν_{CN}), 1155 (m, ρ_{NH}), 1027 (m, ρ_{NH}), 827 (w, ω_{NH}), 765 (w, ω_{NH}), 611 (m, δ_{CN}), 537 (m, δ_{CN}), 433 cm^{-1} (m, τ_{NH}). The detailed crystallographic data for (**1**) is provided in the Supplementary Information.

Hexakis(urea)gallium(III) nitrate (**2**): The compound was synthesized analogously to the preparation of (**1**), (5.56 g, 67%). The product can be recrystallized in a similar fashion like the indium complex. Anal. calcd (%) for $C_6H_{24}GaN_{15}O_{15}$ (616.07 g mol^{-1}) C 11.70 H 3.93 N 34.10; found C 11.51 H 3.92 N 34.07. ^{1}H NMR ([D$_4$] methanol, 25 °C) $\delta = 4.90$ (br, -NH$_2$). ^{13}C{^{1}H} NMR ([D$_4$] methanol, 25 °C): 162.30 (C=O). IR (crystals) $\tilde{\nu} = 3435$ (s, $\nu_{NH,out\ of\ phase}$), 3342 (s, $\nu_{NH\ in\ phase}$), 3235 (s, $\nu_{NH,in\ phase}$), 1630 (s, ν_{CO}), 1560 (s, δ_{NH}), 1505 (s, δ_{NH}), 1332 (s, ν_{CN}), 1152 (m, ρ_{NH}), 1032 (m, ρ_{NH}), 829 (w, ω_{NH}), 762 (w, ω_{NH}), 618 (m, δ_{CN}), 537 (m, δ_{CN}), 433 cm^{-1} (m, τ_{NH}).

Diaqua-tetrakis(urea)zinc(II) nitrate (**3**): Zinc nitrate hexahydrate (5.95 g, 20 mmol) was dissolved in n-butanol. Urea (4.80 g, 80 mmol) was added thereafter into the solution. The solution was stirred for 1 hour and the formation of a visible white precipitated product is observed. The product was washed with butanol to ensure the removal of any by-products and then dried in vacuo (5.90 g, 79%). Crystals for a single crystal structure analysis were obtained by creating a saturated solution in ethanol and layering it with diethyl ether for several days at room temperature. Anal. calcd (%) for $C_4H_{20}N_{10}O_{12}Zn$ (465.64 g mol^{-1}) C 10.32 H 4.33 N 30.08; found C 10.56 H 4.32 N 30.37. ^{1}H NMR ([D$_6$] dimethyl sulfoxide, 25 °C) $\delta = 5.46$ (br, -NH$_2$), 3.47 ppm (br, H$_2$O) ^{13}C{^{1}H} NMR ([D$_6$] dimethyl sulfoxide, 25 °C): 159.62 (C=O). IR (crystals) $\tilde{\nu} = 3447$ (s, $\nu_{NH,out\ of\ phase}$), 3345 (s, $\nu_{NH,in\ phase}$), 3215 (s, $\nu_{NH,in\ phase}$), 1626 (s, ν_{CO}), 1579 (s, δ_{NH}), 1476 (s, δ_{NH}), 1351 (s, ν_{CN}), 1294 (s, ν_{CN}), 1150 (m, ρ_{NH}), 1015 (m, ρ_{NH}), 819 (w, ω_{NH}), 7778 (w, ω_{NH}), 622 (m, δ_{CN}), 591 (m, δ_{CN}), 522 (m, δ_{CN}), 430 cm^{-1} (m, τ_{NH}).

Thin film transistor fabrication

FET substrates were obtained from Fraunhofer IMPS, Dresden. The substrates (15×15 mm^2) consist of a highly n-doped silicon with a 90 nm silicon-oxide dielectric layer. The source-drain electrodes, with a deliberately large channel length L = 20 µm and a channel width W = 10 mm (W/L = 500), were fabricated with 40 nm of gold (interdigital structure) with a 10 nm intermediate adhesion layer of indium tin oxide (ITO) with a device structure in a bottom gate bottom contact device geometry (BGBC). All substrates were sequentially cleaned, with acetone, water and 2-propanol for ten minutes, respectively. The cleaned substrates were exposed for 10 minutes to UV treatment (Nanobio analytics, UVC-1014, 8watts, peak $\lambda = 256$ nm 90%, 184 nm 10%) prior to spin coating, to improve the adhesion of the spin-coating solutions.

For the preparation of the spin coating solutions, stock solutions of the indium(III), gallium(III) and zinc (II) precursor were formed by dissolving five weight % of the respective precursors each in deionised water. After complete dissolution, a clear solution is obtained which is then filtered through a 0.45 µm hydrophylic syringe filter and mixed in the desired ratio (In:Ga:Zn, 7:1:1.5) ratios for spin coating of IGZO thin films. The spin coating was carried out under ambient conditions with a spinning speed of 4000 rpm for 10 seconds. The spin-coated precursor films were subjected to

thermal annealing in ambient atmosphere at desired temperatures for 10 minutes. Repeated iterations of the aforementioned procedure were carried out until desired film thickness was achieved. Finally, the films were annealed for two hours under the respective temperature to generate the functional semiconducting oxide thin films.

Material and device characterization

X-ray diffraction data were collected using a STOE STADI IV 4-circle single crystal diffractometer at ambient temperature. The graphical images of the molecules were generated with using the PLATON software. Transmission electron microscopy (TEM) was carried out using Tecnai F20 (FEI), with an operating voltage of 200 kV. Atomic force microscopy (AFM) was conducted on a MFP-3D™ (Asylum Research) system equipped with silicon cantilevers. (AFM) measurements were carried out with CP-II (Bruker-Veeco), with silicon cantilevers. Thermogravimetric analysis (TGA) measurements were performed using a TG209F1-Iris (Netzsch) thermal analyzer, employing aluminum crucibles. XRD was performed on a Rigaku Miniflex 600@40 kV 15 mA diffractometer using Cu$_{K\alpha1}$ radiation ($\lambda = 1.541$ Å). XPS measurements were performed using a K-Alpha XPS instrument (Thermo Fisher Scientific, East Grinstead, UK) monochromated Al$_{K\alpha}$ X-ray source (30–400 µm spot size). All spectra were referenced to that of hydrocarbon C1s peak at 285.0 eV. TFT characteristics for 8 devices for all temperature ranges measured and were determined with an HP 4155A Semiconductor Parameter Analyzer (Agilent) in a glove box under exclusion of air and moisture in the dark. The fabricated devices were mechanically isolated prior to measurement. Charge carrier mobility (μ_{sat}) and threshold voltage Vth were derived from a linear fitting of the square root of the source-drain current ($\sqrt{I_{ds}}$) as a function of gate-source voltage V_{GS}.

Acknowledgements

S.S., R.H. and J.J.S. acknowledge financial support through the DFG SPP 1569 program. TEM investigations were performed at ERC Jülich under contract ERC-TUD1. We gratefully acknowledge J. Engstler and S. Heinschke (both at TUDa) for performing TEM and XRD analysis.

Conflict of interest

The authors declare no conflict of interest.

Keywords: indium · gallium · metal oxides · molecular precursors · thin film transistors · zinc · urea-nitrate compounds

[1] a) T. Kamiya, H. Hosono, *NPG Asia Mater.* **2010**, *2*, 15; b) E. Fortunato, P. Barquinha, R. Martins, *Adv. Mater.* **2012**, *24*, 2945.

[2] J. F. Wager, B. Yeh, R. L. Hoffman, D. A. Keszler, *Curr. Opin. Solid State Mater. Sci.* **2014**, *18*, 53.

[3] a) M.-G. Kim, M. G. Kanatzidis, A. Facchetti, T. J. Marks, *Nat. Mater.* **2011**, *10*, 382; b) J. W. Hennek, M.-G. Kim, M. G. Kanatzidis, A. Facchetti, T. J. Marks, *J. Am. Chem. Soc.* **2012**, *134*, 9593; c) J. W. Hennek, J. Smith, A. Yan, M.-G. Kim, W. Zhao, V. P. Dravid, A. Facchetti, T. J. Marks, *J. Am. Chem. Soc.* **2013**, *135*, 10729; d) K. Banger, Y. Yamashita, K. Mori, R. Peterson, T. Leedham, J. Rickard, H. Sirringhaus, *Nat. Mater.* **2011**, *10*, 45; e) X. Yu, N. Zhou, J. Smith, H. Lin, K. Stallings, J. Yu, T. J. Marks, A. Fac-

chetti, *ACS Appl. Mater. Interfaces* **2013**, *5*, 7983; f) G. Liu, A. Liu, F. Shan, Y. Meng, B. Shin, E. Fortunato, R. Martins, *Appl. Phys. Lett.* **2014**, *105*, 113509; g) S. Jeong, Y. Jeong, J. Moon, *J. Phys. Chem. C* **2008**, *112*, 11082.

[4] J. S. Chang, A. F. Facchetti, R. Reuss, *IEEE J. Emerging Selected Top. Circ. Syst.* **2017**, *7*, 7.

[5] S. Sanctis, R. C. Hoffmann, M. Bruns, J. J. Schneider, *Adv. Mater. Interfaces* **2018**, 1800324.

[6] S. Sanctis, R. C. Hoffmann, R. Precht, W. Anwand, J. J. Schneider, *J. Mater. Chem. C* **2016**, *4*, 10935.

[7] W. H. Jeong, D. L. Kim, H. J. Kim, *ACS Appl. Mater. Interfaces* **2013**, *5*, 9051.

[8] a) Y. Chen, B. H. Wang, W. Huang, X. A. Zhang, G. Wang, M. J. Leonardi, Y. Huang, Z. Y. Lu, T. J. Marks, A. Facchetti, *Chem. Mater.* **2018**, *30*, 3323; b) R. C. Hoffmann, J. J. Schneider, *Eur. J. Inorg. Chem.* **2014**, 2241.

[9] a) S. T. Aruna, A. S. Mukasyan, *Curr. Opin. Solid State Mater. Sci.* **2008**, *12*, 44; b) A. Varma, A. S. Mukasyan, A. S. Rogachev, K. V. Manukyan, *Chem. Rev.* **2016**, *116*, 14493.

[10] R. Branquinho, D. Salgueiro, L. Santos, P. Barquinha, L. Pereira, R. Martins, E. Fortunato, *ACS Appl. Mater. Interfaces* **2014**, *6*, 19592.

[11] Y. Qiu, L. Gao, *J. Am. Ceram. Soc.* **2004**, *87*, 352.

[12] M. S. Lupin, G. E. Peters, *Thermochim. Acta* **1984**, *73*, 79.

[13] D. G. Tuck, E. J. Woodhouse, P. Carty, *J. Chem. Soc. A* **1966**, DOI: https://doi.org/10.1039/J196600010771077.

[14] a) B. N. Figgis, E. S. Kucharski, J. M. Patrick, A. H. White, *Aust. J. Chem.* **1984**, *37*, 265; b) D. Moon, S. Tanaka, T. Akitsu, J. H. Choi, *Acta Crystallogr. Sect. E* **2015**, *71*, 1336.

[15] H. Aghabozorg, G. J. Palenik, R. C. Stoufer, J. Summers, *Inorg. Chem.* **1982**, *21*, 3903.

[16] T. J. Prior, R. L. Kift, *J. Chem. Crystallogr.* **2009**, *39*, 558.

[17] A. Krawczuk, K. Stadnicka, *Acta Crystallogr. Sect. C* **2007**, *63*, m448.

[18] S. Smeets, M. Lutz, *Acta Crystallogr. Sect. C* **2011**, *67*, m50.

[19] M. Koman, E. Jona, D. Nagy, *Z. Kristallogr.* **1995**, *210*, 873.

[20] a) C.-C. Lin, Y.-Y. Li, *Mater. Chem. Phys.* **2009**, *113*, 334; b) H. M. Ismail, *J. Anal. Appl. Pyrolysis* **1991**, *21*, 315.

[21] T. Arii, A. Kishi, *J. Therm. Anal. Calorim.* **2006**, *83*, 253.

[22] a) V. D. Zhuravlev, V. G. Bamburov, A. R. Beketov, L. A. Perelyaeva, I. V. Baklanova, O. V. Sivtsova, V. G. Vasil'ev, E. V. Vladimirova, V. G. Shevchenko, I. G. Grigorov, *Ceram. Int.* **2013**, *39*, 1379; b) J. Baneshi, M. Haghighi, N. Jodeiri, M. Abdollahifar, H. Ajamein, *Ceram. Int.* **2014**, *40*, 14177.

[23] S. Désilets, P. Brousseau, D. Chamberland, S. Singh, H. Feng, R. Turcotte, K. Armstrong, J. Anderson, *Thermochim. Acta* **2011**, *521*, 59.

[24] Y. Zhao, R. L. Frost, W. N. Martens, *J. Phys. Chem. C* **2007**, *111*, 16290.

[25] H. Y. Playford, A. C. Hannon, E. R. Barney, R. I. Walton, *Chem. Eur. J.* **2013**, *19*, 2803.

[26] A. Nadarajah, M. Z. Wu, K. Archila, M. G. Kast, A. M. Smith, T. H. Chiang, D. A. Keszler, J. F. Wager, S. W. Boettcher, *Chem. Mater.* **2015**, *27*, 5587.

[27] G. H. Kim, B. D. Ahn, H. S. Shin, W. H. Jeong, H. J. Kim, H. J. Kim, *Appl. Phys. Lett.* **2009**, *94*, 233501.

[28] a) B. Cui, L. Zeng, D. Keane, M. J. Bedzyk, D. B. Buchholz, R. P. H. Chang, X. Yu, J. Smith, T. J. Marks, Y. Xia, A. F. Facchetti, J. E. Medvedeva, M. Grayson, *J. Phys. Chem. C* **2016**, *120*, 7467; b) X. Yu, J. Smith, N. Zhou, L. Zeng, P. Guo, Y. Xia, A. Alvarez, S. Aghion, H. Lin, J. Yu, R. P. H. Chang, M. J. Bedzyk, R. Ferragut, T. J. Marks, A. Facchetti, *Proc. Natl. Acad. Sci. USA* **2015**, *112*, 3217.

[29] R. Y. Seung, K. H. Jae, *Phys. Status Solidi A* **2014**, *211*, 2195.

[30] Y.-H. Kim, J.-S. Heo, T.-H. Kim, S. Park, M.-H. Yoon, J. Kim, M. S. Oh, G.-R. Yi, Y.-Y. Noh, S. K. Park, *Nature* **2012**, *489*, 128.

[31] S. Hwang, J. H. Lee, C. H. Woo, J. Y. Lee, H. K. Cho, *Thin Solid Films* **2011**, *519*, 5146.

[32] P. K. Nayak, J. A. Caraveo-Frescas, Z. Wang, M. N. Hedhili, Q. X. Wang, H. N. Alshareef, *Sci. Rep.* **2014**, *4*, 4672.

[33] a) T. S. Kang, K. S. Yoon, G. H. Baek, W. B. Ko, S. M. Yang, B. M. Yeon, J. P. Hong, *Adv. Electron. Mater.* **2017**, *3*, 1600452; b) J. Espinós, J. Morales, A. Barranco, A. Caballero, J. Holgado, A. González-Elipe, *J. Phys. Chem. B* **2002**, *106*, 6921; c) J. W. Na, Y. S. Rim, H. J. Kim, J. H. Lee, S. Hong, H. J. Kim, *ACS Appl. Mater. Interfaces* **2017**, *9*, 29849.

[34] a) R. C. Hoffmann, S. Sanctis, J. J. Schneider, *Inorg. Chem.* **2017**, *56*, 7550; b) N. Zydziak, C. Hübner, M. Bruns, A. P. Vogt, C. Barner-Kowollik, *Polym. Chem.* **2013**, *4*, 1525; c) N. Zydziak, C. M. Preuss, V. Winkler, M. Bruns, C. Hübner, C. Barner-Kowollik, *Macromol. Rapid Commun.* **2013**, *34*, 672.

[35] a) S. Jeong, Y.-G. Ha, J. Moon, A. Facchetti, T. J. Marks, *Adv. Mater.* **2010**, *22*, 1346; b) H. Hideo, *SID Symp. Dig. Tech. Pap.* **2007**, *38*, 1830.

[36] R. A. John, N. A. Chien, S. Shukla, N. Tiwari, C. Shi, N. G. Ing, N. Mathews, *Chem. Mater.* **2016**, *28*, 8305.

[37] CCDC 1863798 (1) and 704906 (3) contain the supplementary crystallographic data for this paper. These data can be obtained free of charge from The Cambridge Crystallographic Data Centre.

Manuscript received: September 12, 2018
Revised manuscript received: October 7, 2018
Version of record online: November 13, 2018

108

The investigations performed within this book demonstrate the synthesis and full structural elucidation of novel single-source molecular precursors for the formation of solution-processed binary and multinary metal oxide dielectrics and semiconductors as well as their integration in capacitor and thin-film transistor devices.

In the course of this book, combustible malonato complexes of aluminum and yttrium were synthesized by performing the systematic introduction of reactive nitro groups into the diethyl malonate ligand framework, which enhances the exothermic decomposition behavior of the precursors, due to these functional nitro groups. Consequently, the employment of the molecular precursors tris[(diethyl-2-nitromalonato)] aluminum(III) (Al-DEM-NO$_2$) and bis(diethyl-2-nitromalonato) nitrato yttrium(III) (Y-DEM-NO$_2$) enable the synthesis of electrically well-performing dielectric Al$_x$O$_y$ and Y$_x$O$_y$ thin films at PDA temperatures as low as 250 °C, without the requirement of any additive. Processing temperatures of 350 °C result in capacitor devices exhibiting excellent dielectric properties. The enhancement of the dielectric properties at elevated temperatures, corresponds to the progressed conversion of the respective functional metal oxide, which was confirmed by XPS. Finally, solution-processed (350 °C) metal oxide Al$_x$O$_y$ and Y$_x$O$_y$ thin films could demonstrate their potential as gate dielectric in thin-film transistor devices, using indium zinc oxide (IZO) as active semiconductor and exhibiting reasonable TFT characteristics.

Furthermore, the employment of the malonato complexes enables the solution-based synthesis of the amorphous ternary metal oxide dielectric yttrium aluminium oxide (YAl$_x$O$_y$). Such hybrid composites represent an effective approach to optimize the dielectric properties by combining a material possessing a high dielectric constant (Y$_x$O$_y$) with another one providing a wide band gap (Al$_x$O$_y$) and thus allowing the electrical fine-tuning of the gate dielectric. The yttrium inclusion into the Al$_x$O$_y$ host lattice reaches a saturation at an incorporation of 30 mol-% (30-YAl$_x$O$_y$), which is corresponding to a reduced amount of carbonate species and thus resulting in a decreased leakage current density of the capacitor devices. As a result, the 30-YAl$_x$O$_y$ composition displays optimum overall dielectric properties, exhibiting a high capacity with almost no frequency dispersion, high electrical breakdown fields and low leakage current densities. Consequently, the integration of the 30-YAl$_x$O$_y$ dielectric leads to TFT devices, exhibiting decent electrical performance characteristics.

Besides, the malonato complexes depict strong absorptions at wavelength $\lambda < 250$ nm and thus we investigated their potential towards low-temperature (150 °C) DUV-mediated solution-processing of YAl_xO_y dielectrics. Unfortunately, the decomposition of the Y-DEM-NO$_2$ precursor could not be initiated even under irradiation of $\lambda = 160$ nm, resulting in degraded YAl_xO_y dielectrics. In contrast, the photolytic decomposition of Al-DEM-NO$_2$ ($\lambda = 160$ nm) leads to well-performing Al_xO_y dielectric thin films, compatible for the fabrication of flexible electronics. Furthermore, the utilization of such malonato complexes could be transferred to various metal cations, enabling a possible photoactivation of the respective metal oxide layers for the application in e.g solar cells[249-253] or gas sensors.[254-256]

Another focus of this work, was the optimization of the combustion synthesis of metal oxides based on a "fuel/oxidizer" reaction by directly combining the "oxidizer" (metal nitrates) and the "fuel" (urea ligands) in one defined molecule. This strategy allows an eco-friendly and direct solution-processing of the respective metal oxides. Furthermore, the urea-nitrate precursors are highly soluble in water and undergo a clean thermal decomposition, releasing various volatile gaseous by-products. As a result, a higher degree of reproducibility can be achieved, which enables a more systematic study of the respective metal oxide formation. Thus, we performed the aqueous combustion synthesis of dielectric aluminum oxide and yttrium oxide thin films, by introducing the defined urea nitrate coordination compounds with the compositions $[Al(CH_4N_2O)_6](NO_3)_3$ and $[Y(CH_4N_2O)_4(NO_3)_2](NO_3)_2$. The fabrication of Al_xO_y based capacitors, exhibit remarkably high areal capacities and very low current leakage densities from 250 °C onwards, while Y_xO_y thin films only start to perform at processing temperatures $T \geq 300$ °C and display an increased tendency of short circuits, due to an enhanced degree of crystallinity. Additionally, we extended this approach to the formation of quaternary amorphous semiconducting IGZO thin films, by reporting on the synthesis and full structural characterization of well-defined urea nitrate coordination compounds of indium, gallium and zinc. TFTs were fabricated from a defined precursor ratio of (In:Ga:Zn, 7:1:1.5) and processed at 300 °C, exhibiting good electrical performance characteristics with respect to the field effect mobility, on/off current ratio and threshold voltage. Furthermore, an active transistor performance can be achieved at processing temperatures as low as 200 °C, by employing this aqueous solution-processed combustion route.

Besides, a novel strategy for the formation of zinc tin oxide (ZTO) was investigated, due to the future need of indium-free, amorphous oxide semiconductor materials. In fact, the choice of the tin precursor had been mainly restricted to tin (II) chloride, which requires high decomposition temperatures and suffers from Cl⁻ ion impurities found to be present within the final metal oxide thin films, increasing the electron trap density and thus deteriorating the TFT performance. Consequently, we developed a molecular precursor approach by employing the novel precursor compound bis[(methoxyimino)-propanoato]tin(II) (Sn-oximato) as well as the established zinc oximato precursor, enabling the realization of lower processing temperatures in comparison to the frequently used $SnCl_2$ precursor. Furthermore, the utilization of the oximato precursors ensures a similar thermal decomposition behavior of the precursors, resulting in a uniform distribution of the Zn, Sn(IV) and O species across the surface and in the depth of the thin films, confirmed by auger electron spectroscopy. Finally, the employment of the oximato precursors, allows the fabrication of ZTO based TFT devices, whereby a precursor ratio of Sn:Zn = 7:3 yields the optimum TFT performance characteristics. The requirement for a higher Sn ratio is due to a higher defect concentration of SnO_2 in comparison to ZnO, derived from the oximato precursors, which is in contrast to the conventional precursor ratio of Sn:Zn = 3:7, employing metal chlorides.

In conclusion, novel single-source molecular precursors were synthesized and characterized in detail in the course of this book, in order to overcome existing obstacles regarding the formation of solution-processed high-*k* dielectrics as well as multinary metal oxide semiconductors. The precursor chemistry, material analysis as well as the electronic performance of low-temperature solution-processed functional metal oxides was performed, by introducing combustible malonato and urea-nitrate complexes as well as molecular oximato precursors. Finally, the presented work within this book certainly possess great potential towards the improvement of precursor-based solution routes and material deposition for the formation of binary and multinary gate dielectrics as well as metal oxide semiconductors towards the application in e.g. TFT devices, for future flexible electronics.

8 Appendix

This section of the book contains all the necessary reproduction rights and permissions for the first- and co-author publications as well as figures obtained from third-party publications, which are presented herein.

Synthesis, dielectric properties and application in a thin film transistor device of amorphous aluminum oxide Al_xO_y using a molecular based precursor route

N. Koslowski, S. Sanctis, R. C. Hoffmann, M. Bruns and J. J. Schneider, *J. Mater. Chem. C*, 2019, **7**, 1048
DOI: 10.1039/C8TC04660C

Section 6.1 Synthesis, dielectric properties and application in a thin film transistor device of amorphous aluminum oxide Al_xO_y using a molecular based precursor route.

Synthesis, oxide formation, properties and thin film transistor properties of yttrium and aluminium oxide thin films employing a molecular-based precursor route

N. Koslowski, R. C. Hoffmann, V. Trouillet, M. Bruns, S. Foro and J. J. Schneider, *RSC Adv.*, 2019, **9**, 31386

DOI: 10.1039/C9RA05348D

Section 6.2 Synthesis, oxide formation, properties and thin film transistor properties of yttrium and aluminium oxide thin films employing a molecular-based precursor route.

Solution-processed amorphous yttrium aluminium oxide YAl$_x$O$_y$ and aluminum oxide Al$_x$O$_y$ and their functional dielectric properties and performance in thin-film transistors

N. Koslowski, V. Trouillet and J. J. Schneider, *J. Mater. Chem. C*, 2020, **8**, 8521

DOI: 10.1039/D0TC01876G

If you are not the author of this article and you wish to reproduce material from it in a third party non-RSC publication you must formally request permission using Copyright Clearance Center. Go to our Instructions for using Copyright Clearance Center page for details.

Authors contributing to RSC publications (journal articles, books or book chapters) do not need to formally request permission to reproduce material contained in this article provided that the correct acknowledgement is given with the reproduced material.

Reproduced material should be attributed as follows:

- For reproduction of material from NJC:
  Reproduced from Ref. XX with permission from the Centre National de la Recherche Scientifique (CNRS) and The Royal Society of Chemistry.
- For reproduction of material from PCCP:
  Reproduced from Ref. XX with permission from the PCCP Owner Societies.
- For reproduction of material from PPS:
  Reproduced from Ref. XX with permission from the European Society for Photobiology, the European Photochemistry Association, and The Royal Society of Chemistry.
- For reproduction of material from all other RSC journals and books:
  Reproduced from Ref. XX with permission from The Royal Society of Chemistry.

If the material has been adapted instead of reproduced from the original RSC publication "Reproduced from" can be substituted with "Adapted from".

In all cases the Ref. XX is the XXth reference in the list of references.

If you are the author of this article you do not need to formally request permission to reproduce figures, diagrams etc. contained in this article in third party publications or in a thesis or dissertation provided that the correct acknowledgement is given with the reproduced material.

Reproduced material should be attributed as follows:

- For reproduction of material from NJC:
  [Original citation] - Reproduced by permission of The Royal Society of Chemistry (RSC) on behalf of the Centre National de la Recherche Scientifique (CNRS) and the RSC
- For reproduction of material from PCCP:
  [Original citation] - Reproduced by permission of the PCCP Owner Societies
- For reproduction of material from PPS:
  [Original citation] - Reproduced by permission of The Royal Society of Chemistry (RSC) on behalf of the European Society for Photobiology, the European Photochemistry Association, and RSC
- For reproduction of material from all other RSC journals:
  [Original citation] - Reproduced by permission of The Royal Society of Chemistry

If you are the author of this article you still need to obtain permission to reproduce the whole article in a third party publication with the exception of reproduction of the whole article in a thesis or dissertation.

Section 6.3 Solution-processed amorphous yttrium aluminium oxide YAl$_x$O$_y$ and aluminum oxide Al$_x$O$_y$ and their functional dielectric properties and performance in thin-film transistors.

Section 6.4 Toward an understanding of thin-film transistor performance in solution-processed amorphous zinc tin oxide (ZTO) thin films.

Aqueous Solution Processing of Combustible Precursor Compounds into Amorphous Indium Gallium Zinc Oxide (IGZO) Semiconductors for Thin Film Transistor Applications

Author: Shawn Sanctis, Rudolf C. Hoffmann, Nico Koslowski, et al

Publication: Chemistry - An Asian Journal

Publisher: John Wiley and Sons

Date: Nov 13, 2018

© 2018 Wiley-VCH Verlag GmbH & Co. KGaA, Weinheim

Order Completed

Thank you for your order.

This Agreement between Mr. Nico Koslowski ("You") and John Wiley and Sons ("John Wiley and Sons") consists of your license details and the terms and conditions provided by John Wiley and Sons and Copyright Clearance Center.

Your confirmation email will contain your order number for future reference.

License Number	4920370087012		🖶 Printable Details
License date	Oct 01, 2020		

📋 Licensed Content

Licensed Content Publisher	John Wiley and Sons
Licensed Content Publication	Chemistry - An Asian Journal
Licensed Content Title	Aqueous Solution Processing of Combustible Precursor Compounds into Amorphous Indium Gallium Zinc Oxide (IGZO) Semiconductors for Thin Film Transistor Applications
Licensed Content Author	Shawn Sanctis, Rudolf C. Hoffmann, Nico Koslowski, et al
Licensed Content Date	Nov 13, 2018
Licensed Content Volume	13
Licensed Content Issue	24
Licensed Content Pages	8

🖥 Order Details

Type of use	Dissertation/Thesis
Requestor type	Author of this Wiley article
Format	Print and electronic
Portion	Full article
Will you be translating?	No

📗 About Your Work

Title	Ph.D.
Institution name	Technische Universität Darmstadt
Expected presentation date	Dec 2020

🗁 Additional Data

Section 6.5 Aqueous solution processing of combustible precursor compounds into amorphous indium gallium zinc oxide (IGZO) semiconductors for thin film transistor applications.

High-K materials and metal gates for CMOS applications

Author: John Robertson,Robert M. Wallace

Publication: Materials Science and Engineering: R: Reports

Publisher: Elsevier

Date: February 2015

Copyright © 2014 Elsevier B.V. All rights reserved.

Order Completed

Thank you for your order.

This Agreement between Mr. Nico Koslowski ("You") and Elsevier ("Elsevier") consists of your license details and the terms and conditions provided by Elsevier and Copyright Clearance Center.

Your confirmation email will contain your order number for future reference.

License Number	4921851358550		🖶 Printable Details
License date	Oct 04, 2020		

☑ Licensed Content

Licensed Content Publisher	Elsevier
Licensed Content Publication	Materials Science and Engineering: R: Reports
Licensed Content Title	High-K materials and metal gates for CMOS applications
Licensed Content Author	John Robertson,Robert M. Wallace
Licensed Content Date	Feb 1, 2015
Licensed Content Volume	88
Licensed Content Issue	n/a
Licensed Content Pages	41
Journal Type	S&T

🗓 Order Details

Type of Use	reuse in a thesis/dissertation
Portion	figures/tables/illustrations
Number of figures/tables /illustrations	1
Format	both print and electronic
Are you the author of this Elsevier article?	No
Will you be translating?	No

📄 About Your Work

Title	Ph.D.
Institution name	Technische Universität Darmstadt
Expected presentation date	Dec 2020

🗁 Additional Data

Portions	Figure 2

Figure 2 Leakage current density vs. voltage for various layer thicknesses of SiO_2. Measurements from[38].

Copyright Clearance Center Marketplace™

Royal Society of Chemistry - License Terms and Conditions

Order Date	06-Dec-2020	**Type of Use**	Republish in a thesis/dissertation
Order license ID	1082164-2	**Publisher**	ROYAL SOCIETY OF CHEMISTRY,
ISSN	1364-5501	**Portion**	Image/photo/illustration

LICENSED CONTENT

Publication Title	Journal of materials chemistry	**Country**	United Kingdom of Great Britain and Northern Ireland
Author/Editor	Royal Society of Chemistry (Great Britain)	**Rightsholder**	Royal Society of Chemistry
Date	01/01/1991	**Publication Type**	e-Journal
Language	English	**URL**	http://firstsearch.oclc.org

REQUEST DETAILS

Portion Type	Image/photo/illustration	**Distribution**	Worldwide
Number of images / photos / illustrations	1	**Translation**	Original language of publication
Format (select all that apply)	Print, Electronic	**Copies for the disabled?**	No
Who will republish the content?	Publisher, not-for-profit	**Minor editing privileges?**	No
Duration of Use	Life of current edition	**Incidental promotional use?**	No
Lifetime Unit Quantity	Up to 499	**Currency**	EUR
Rights Requested	Main product		

NEW WORK DETAILS

Title	Ph.D.	**Institution name**	TU Darmstadt
Instructor name	Prof. Dr. Jörg J. Schneider	**Expected presentation date**	2020-12-21

ADDITIONAL DETAILS

Order reference number	N/A	**The requesting person / organization to appear on the license**	Nico Koslowski

REUSE CONTENT DETAILS

Title, description or numeric reference of the portion(s)	Scheme 2	**Title of the article/chapter the portion is from**	3.1 Detailed mechanism of precursor decomposition
Editor of portion(s)	N/A		

Figure 4 (right side) decomposition mechanism of zinc oximato complexes, determined by TG-MS and TG-IR.

Dear Nico Koslowski,

Thank you for your request to reuse JSAP material *in your thesis,* **"Approaching metal oxide high-k dielectrics and semiconductors by solution-processing of molecular precursors"**

Regarding:

- ***Figure 2*** *from* **"Review of solution-processed oxide thin-film transistors"**

We are happy to grant permission for the use you request, provided:

- The Copyright Policy of JSAP is followed https://www.jsap.or.jp/english/copyright-policy
- The article is cited as a reference
- The copyright line including year of publication is added to the reproduced content (at the end of the figure captions):
    - Copyright (2014) The Japan Society of Applied Physics

This does not apply to any material/figure which is credited to another source in the publication or has been obtained from a third party. Express permission for such materials/figures must be obtained from the copyright owner.

Kind regards,

Sophie
Copyright & Permissions Team
Sophie Brittain - Rights & Permissions Assistant
Cameron Wood - Legal & Rights Adviser
Contact Details
E-mail: permissions@ioppublishing.org

For further information about copyright and how to request permission:
https://publishingsupport.iopscience.iop.org/copyright-journals/
See also: https://publishingsupport.iopscience.iop.org/
Please see our Author Rights Policy https://publishingsupport.iopscience.iop.org/author-rights-policies/
Please note: We do not provide signed permission forms as a separate attachment. Please print this email and provide it to your publisher as proof of permission. **Please note**: Any statements made by IOP Publishing to the effect that authors do not need to get permission to use any content where IOP Publishing is not the publisher is not intended to constitute any sort of legal advice. Authors must make their own decisions as to the suitability of the content they are using and whether they require permission for it to be published within their article.

Figure 5 Percentage of publication amount in dependency of the investigated semiconductor material in the years 2008-2013.

Figure 7 Schematic illustration of parallel-plate capacitors (a) without and (b) with dielectric layer. (c) Illustrated dipole polarization with (top) and without (bottom) an external electrical bias.

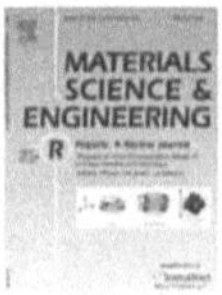

Figure 8 Schematic illustration of the dielectric polarization mechanisms: (a) electronic polarization, (b) ionic polarization, (c) orientation polarization. (d) Contribution of electronic, ionic, and orientation polarization regarding the dielectric constant at various frequencies.

Figure 9 Schematic illustration of the Poole-Frenkel effect, a) change of the trap geometry at the transition from equilibrium to the interaction with an external electric field, b) reduction of the potential barrier caused by the external electric field, which increases the probability of the electron to promote from the trapped state to the conduction band.

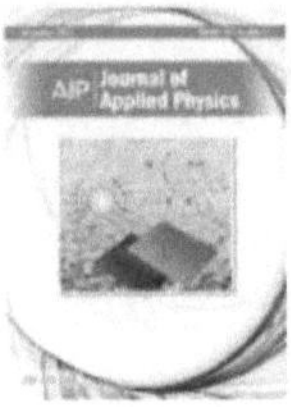

Figure 10 Schematic illustration of the mechanism of a) the Schottky emission in comparison to b) the Poole-Frenkel emission for an Al_2O_3 dielectric, using ITO as gate-electrode and Al as top-electrode.

Figure 13 TFT device geometries: (a) staggered, bottom-gate top-contact (BGTC), (b) staggered, top-gate bottom-contact (TGBC), (c) coplanar, bottom-gate bottom-contact (BGBC) and, (d) coplanar, top-gate top-contact (TGTC).

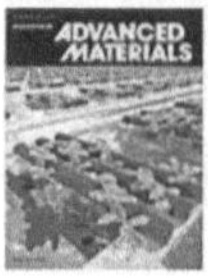

Figure 14 Electrical characterization of a thin-film transistor (a) output curves (I_D against V_D) and (b) transfer curve (I_D against V_G).

Solution Processed Metal Oxide High-κ Dielectrics for Emerging Transistors and Circuits

Author: Ao Liu, Huihui Zhu, Huabin Sun, et al

Publication: Advanced Materials

Publisher: John Wiley and Sons

Date: Jun 14, 2018

© 2018 WILEY-VCH Verlag GmbH & Co. KGaA, Weinheim

Order Completed

Thank you for your order.

This Agreement between Mr. Nico Koslowski ("You") and John Wiley and Sons ("John Wiley and Sons") consists of your license details and the terms and conditions provided by John Wiley and Sons and Copyright Clearance Center.

Your confirmation email will contain your order number for future reference.

License Number	4921400972974
License date	Oct 03, 2020

🖶 Printable Details

🗹 Licensed Content

Licensed Content Publisher	John Wiley and Sons
Licensed Content Publication	Advanced Materials
Licensed Content Title	Solution Processed Metal Oxide High-κ Dielectrics for Emerging Transistors and Circuits
Licensed Content Author	Ao Liu, Huihui Zhu, Huabin Sun, et al
Licensed Content Date	Jun 14, 2018
Licensed Content Volume	30
Licensed Content Issue	33
Licensed Content Pages	39

🗎 Order Details

Type of use	Dissertation/Thesis
Requestor type	University/Academic
Format	Print and electronic
Portion	Figure/table
Number of figures/tables	2
Will you be translating?	No

🗎 About Your Work

Title	Ph.D.
Institution name	Technische Universität Darmstadt
Expected presentation date	Dec 2020

🗁 Additional Data

Portions	Figure 4 and Figure 6b)

Figure 15 Periodic table of the elements. The binary oxides with higher k values than SiO_2 are highlighted; the color scale indicates the amount of the dielectric constant.

Figure 16 Dielectric constant (k) and band gap energy (E_G) of various binary oxide dielectrics.

Royal Society of Chemistry - License Terms and Conditions

This is a License Agreement between Nico Koslowski ("You") and Royal Society of Chemistry ("Publisher") provided by Copyright Clearance Center ("CCC"). The license consists of your order details, the terms and conditions provided by Royal Society of Chemistry, and the CCC terms and conditions.

All payments must be made in full to CCC.

Order Date	04-Oct-2020	Type of Use	Republish in a thesis/dissertation
Order license ID	1067624-1	Publisher	ROYAL SOCIETY OF CHEMISTRY
ISSN	1460-4744	Portion	Chart/graph/table/figure

LICENSED CONTENT

Publication Title	Chemical Society reviews	Country	United Kingdom of Great Britain and Northern Ireland
Author/Editor	Royal Society of Chemistry (Great Britain)		
Date	01/01/1972	Rightsholder	Royal Society of Chemistry
Language	English	Publication Type	e-Journal
		URL	http://www.rsc.org/csr

REQUEST DETAILS

Portion Type	Chart/graph/table/figure	Distribution	Worldwide
Number of charts / graphs / tables / figures requested	1	Translation	Original language of publication
		Copies for the disabled?	No
Format (select all that apply)	Print, Electronic	Minor editing privileges?	No
Who will republish the content?	Publisher, not-for-profit	Incidental promotional use?	No
Duration of Use	Life of current edition	Currency	EUR
Lifetime Unit Quantity	Up to 499		
Rights Requested	Main product		

NEW WORK DETAILS

Title	Ph.D.	Institution name	TU Darmstadt
Instructor name	Prof. Dr. Jörg J. Schneider	Expected presentation date	2020-12-14

ADDITIONAL DETAILS

Order reference number	N/A	The requesting person / organization to appear on the license	Nico Koslowski

Figure 18 Schematic illustration of various solution-based deposition techniques.

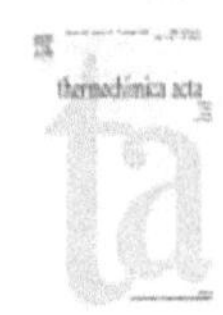

Analyses of the thermal decomposition of urea nitrate at high temperature

Author:
Sylvain Désilets,Patrick Brousseau,Daniel Chamberland,Shanti Singh,Hongtu Feng,Richard Turcotte,Kelly Armstrong,John Anderson

Publication: Thermochimica Acta

Publisher: Elsevier

Date: 10 July 2011

Order Completed

Thank you for your order.

This Agreement between Mr. Nico Koslowski ("You") and Elsevier ("Elsevier") consists of your license details and the terms and conditions provided by Elsevier and Copyright Clearance Center.

Your confirmation email will contain your order number for future reference.

License Number	4963090703430	🖶 Printable Details
License date	Dec 06, 2020	

📑 Licensed Content

Licensed Content Publisher	Elsevier
Licensed Content Publication	Thermochimica Acta
Licensed Content Title	Analyses of the thermal decomposition of urea nitrate at high temperature
Licensed Content Author	Sylvain Désilets,Patrick Brousseau,Daniel Chamberland,Shanti Singh,Hongtu Feng,Richard Turcotte,Kelly Armstrong,John Anderson
Licensed Content Date	Jul 10, 2011
Licensed Content Volume	521
Licensed Content Issue	1-2
Licensed Content Pages	7
Journal Type	S&T

📋 Order Details

Type of Use	reuse in a thesis/dissertation
Portion	figures/tables/illustrations
Number of figures/tables /illustrations	1
Format	both print and electronic
Are you the author of this Elsevier article?	No
Will you be translating?	No

📄 About Your Work

Title	Ph.D.
Institution name	Technische Universität Darmstadt
Expected presentation date	Dec 2020

🗂 Additional Data

Portions	Fig. 8

Figure 24 (right side): Decomposition mechanism of urea-nitrate from 25 °C to 360 °C, determined by TG-FTIR-MS.

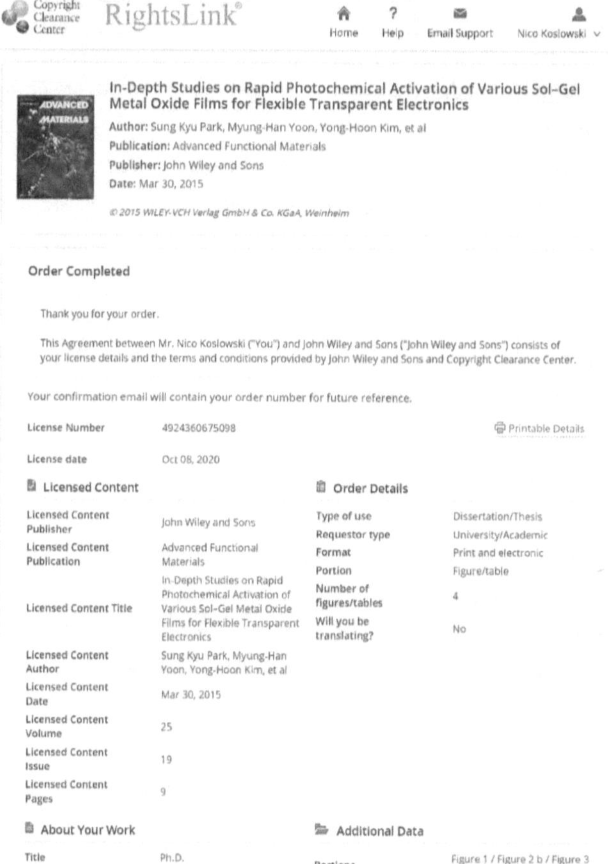

Figure 28 a) Schematic illustration of the general steps involved in the low-temperature photoactivation of various sol–gel metal oxide films; b) Mechanisms of the photoactivated solution process, proposed by Park et al.

Figure 29 a) FTIR spectra of Al_xO_y thin films processed at 150 °C; without (blue line) and with (red line) DUV treatment, recorded at various UV exposure times. b) O 1s XPS spectra of Al_xO_y thin films processed at different conditions: (A) treated at 100 °C for 5 min (B) without, (C) with DUV irradiation, and (D) thermally annealed at 350 °C for one hour; b) Film thicknesses of Al_xO_y thin films processed at various temperatures; without (blue) and with (red) DUV-irradiation.

Figure 30 a) Leakage current density vs. e-field curves and b) areal capacitance vs. frequency curves of Al_xO_y dielectrics processed without DUV irradiation, with DUV irradiation and at a high annealing temperature.

9 References

1. A. Facchetti and T. Marks, *Transparent electronics: from synthesis to applications*, John Wiley & Sons, 2010.

2. P. K. Nayak, M. N. Hedhili, D. Cha and H. N. Alshareef, *Appl. Phys. Lett.*, 2013, **103**, 033518.

3. R. C. Frunză, B. Kmet, M. Jankovec, M. Topič and B. Malič, *Mater. Res. Bull.*, 2014, **50**, 323-328.

4. Y. B. Yoo, J. H. Park, K. H. Lee, H. W. Lee, K. M. Song, S. J. Lee and H. K. Baik, *J. Mater. Chem. C*, 2013, **1**, 1651-1658.

5. C.-G. Lee and A. Dodabalapur, *J. Electron. Mater.*, 2012, **41**, 895-898.

6. A. Liu, G. X. Liu, H. H. Zhu, F. Xu, E. Fortunato, R. Martins and F. K. Shan, *ACS Appl. Mater. Interfaces*, 2014, **6**, 17364-17369.

7. L. Zhu, G. He, J. Lv, E. Fortunato and R. Martins, *RSC Adv.*, 2018, **8**, 16788-16799.

8. C.-Y. Tsay, C.-H. Cheng and Y.-W. Wang, *Ceram. Int.*, 2012, **38**, 1677-1682.

9. K. Song, W. Yang, Y. Jung, S. Jeong and J. Moon, *J. Mater. Chem.*, 2012, **22**, 21265-21271.

10. A. Liu, G. Liu, H. Zhu, Y. Meng, H. Song, B. Shin, E. Fortunato, R. Martins and F. Shan, *Curr. Appl. Phys.*, 2015, **15**, S75-S81.

11. F. Xu, A. Liu, G. Liu, B. Shin and F. Shan, *Ceram. Int.*, 2015, **41**, S337-S343.

12. G. Adamopoulos, S. Thomas, D. D. C. Bradley, M. A. McLachlan and T. D. Anthopoulos, *Appl. Phys. Lett.*, 2011, **98**, 123503.

13. C. Avis and J. Jang, *J. Mater. Chem.*, 2011, **21**, 10649-10652.

14. A. Liu, G. Liu, H. Zhu, B. Shin, E. Fortunato, R. Martins and F. Shan, *RSC Adv.*, 2015, **5**, 86606-86613.

15. W. Xu, H. Wang, F. Xie, J. Chen, H. Cao and J.-B. Xu, *ACS Appl. Mater. Interfaces*, 2015, **7**, 5803-5810.

16. W. Xu, H. Li, J.-B. Xu and L. Wang, *ACS Appl. Mater. Interfaces*, 2018, **10**, 25878-25901.

17. S. Park, C.-H. Kim, W.-J. Lee, S. Sung and M.-H. Yoon, *Mater. Sci. Eng., R*, 2017, **114**, 1-22.

18. R. Branquinho, D. Salgueiro, L. Santos, P. Barquinha, L. Pereira, R. Martins and E. Fortunato, *ACS Appl. Mater. Interfaces*, 2014, **6**, 19592-19599.

19. K.-H. Lim, J. Lee, J.-E. Huh, J. Park, J.-H. Lee, S.-E. Lee and Y. S. Kim, *J. Mater. Chem. C*, 2017, **5**, 7768-7776.

20. S. R. Thomas, P. Pattanasattayavong and T. D. Anthopoulos, *Chem. Soc. Rev.*, 2013, **42**, 6910-6923.

21. Y. Kuo, *Electrochem. Soc. Interface*, 2013, **22**, 55.

22. J.-P. Moy, *Thin Solid Films*, 1999, **337**, 213-221.

23. S. O. Kasap and J. A. Rowlands, *J. Mater. Sci.: Mater. Electron.*, 2000, **11**, 179-198.

24. F. Liao, C. Chen and V. Subramanian, *Sens. Actuator A Phys.*, 2005, **107**, 849-855.

25. P. Estrela and P. Migliorato, *J. Mater. Chem.*, 2007, **17**, 219-224.

26. R. H. Reuss, B. R. Chalamala, A. Moussessian, M. G. Kane, A. Kumar, D. C. Zhang, J. A. Rogers, M. Hatalis, D. Temple and G. Moddel, *Proc. IEEE*, 2005, **93**, 1239-1256.

27. K. Nomura, H. Ohta, A. Takagi, T. Kamiya, M. Hirano and H. Hosono, *nature*, 2004, **432**, 488-492.

28. T. D. Anthopoulos, Y. Y. Noh and O. D. Jurchescu, *Adv. Funct. Mater.*, 2020, **30**, 2001678.

29. A. Tixier-Mita, S. Ihida, B.-D. Ségard, G. A. Cathcart, T. Takahashi, H. Fujita and H. Toshiyoshi, *Jpn. J. Appl. Phys.*, 2016, **55**, 04EA08.

30. J.-Y. Kwon, D.-J. Lee and K.-B. Kim, *Electron. Mater. Lett.*, 2011, **7**, 1-11.

31. W. Xu, M. Dai, L. Liang, Z. Liu, X. Sun, Q. Wan and H. Cao, *J. Phys. D Appl. Phys.*, 2012, **45**, 205103.

32. P. Barquinha, R. Martins, L. Pereira and E. Fortunato, *Transparent oxide electronics: from materials to devices*, John Wiley & Sons, 2012.

33. R. G. Gordon, J. Becker, D. Hausmann and S. Suh, *Chem. Mater.*, 2001, **13**, 2463-2464.

34. J. Robertson and R. M. Wallace, *Mater. Sci. Eng., R*, 2015, **88**, 1-41.

35. R. M. Wallace and G. Wilk, *MRS Bull.*, 2002, **27**, 186-191.

36. J. Robertson, *EPJ Appl. Phys*, 2004, **28**, 265-291.

37. M. Hirose, M. Koh, W. Mizubayashi, H. Murakami, K. Shibahara and S. Miyazaki, *Semicond. Sci. Technol.*, 2000, **15**, 485.

38. S.-H. Lo, D. Buchanan, Y. Taur and W. Wang, *IEEE Electron Device Lett.*, 1997, **18**, 209-211.

39. Y.-J. Cho, J.-H. Shin, S. Bobade, Y.-B. Kim and D.-K. Choi, *Thin Solid Films*, 2009, **517**, 4115-4118.

40. Y. S. Chun, S. Chang and S. Y. Lee, *Microelectron. Eng.*, 2011, **88**, 1590-1593.

41. T. Kamiya and H. Hosono, *NPG Asia Mater.*, 2010, **2**, 15-22.

42. J. Robertson, *Rep. Prog. Phys.*, 2005, **69**, 327.

43. M. D. Groner, J. W. Elam, F. H. Fabreguette and S. M. George, *Thin Solid Films*, 2002, **413**, 186-197.

44. J. Kim, D. Kwon, K. Chakrabarti, C. Lee, K. Oh and J. Lee, *J. Appl. Phys.*, 2002, **92**, 6739-6742.

45. M. S. Oh, K. Lee, J. Song, B. H. Lee, M. M. Sung, D. Hwang and S. Im, *J. Electrochem. Soc.*, 2008, **155**, H1009.

46. M. Esposto, S. Krishnamoorthy, D. N. Nath, S. Bajaj, T.-H. Hung and S. Rajan, *Appl. Phys. Lett.*, 2011, **99**, 133503.

47. A. Illiberi, B. Cobb, A. Sharma, T. Grehl, H. Brongersma, F. Roozeboom, G. Gelinck and P. Poodt, *ACS Appl. Mater. Interfaces*, 2015, **7**, 3671-3675.

48. S.-M. Yoon, N.-J. Seong, K. Choi, G.-H. Seo and W.-C. Shin, *ACS Appl. Mater. Interfaces*, 2017, **9**, 22676-22684.

49. T. Maruyama and T. Nakai, *Appl. Phys. Lett.*, 1991, **58**, 2079-2080.

50. T. M. Klein, D. Niu, W. S. Epling, W. Li, D. M. Maher, C. C. Hobbs, R. I. Hegde, I. J. R. Baumvol and G. N. Parsons, *Appl. Phys. Lett.*, 1999, **75**, 4001-4003.

51. A. N. Gleizes, C. Vahlas, M. M. Sovar, D. Samélor and M. C. Lafont, *Chem. Vap. Depos.*, 2007, **13**, 23-29.

52. D. B. Potter, M. J. Powell, I. P. Parkin and C. J. Carmalt, *J. Mater. Chem. C*, 2018, **6**, 588-597.

53. T. Seino and T. Sato, *J. Vac. Sci. Technol. A*, 2002, **20**, 634-637.

54. Y. S. Jung, J. Y. Seo, D. W. Lee and D. Y. Jeon, *Thin Solid Films*, 2003, **445**, 63-71.

55. X. Tang, F. Luo, F. Ou, W. Zhou, D. Zhu and Z. Huang, *Appl. Surf. Sci.*, 2012, **259**, 448-453.

56. P. S, M. Gowravaram, J. S, M. Kannan and G. V, *Thin Solid Films*, 2012, **520**, 2689–2694.

57. W. H. Ha, M. H. Choo and S. Im, *J. Non Cryst. Solids*, 2002, **303**, 78-82.

58. M. Voigt and M. Sokolowski, *Mater. Sci. Eng. B*, 2004, **109**, 99-103.

59. H. Yabuta, M. Sano, K. Abe, T. Aiba, T. Den, H. Kumomi, K. Nomura, T. Kamiya and H. Hosono, *Appl. Phys. Lett.*, 2006, **89**, 112123.

60. D. Redinger and V. Subramanian, *IEEE Trans. Electron. Devices*, 2007, **54**, 1301-1307.

61. H.-C. Cheng, C.-F. Chen and C.-C. Lee, *Thin Solid Films*, 2006, **498**, 142-145.

62. P. Gogoi, R. Saikia and S. Changmai, *J. Semicond.*, 2015, **36**, 044002.

63. W. Xu, H. Wang, L. Ye and J. Xu, *J. Mater. Chem. C*, 2014, **2**, 5389-5396.

64. B. Norris, J. Anderson, J. Wager and D. Keszler, *J. Phys. D Appl. Phys.*, 2003, **36**, L105.

65. J. Lee, H. Seul and J. K. Jeong, *J. Alloys Compd.*, 2018, **741**, 1021-1029.

66. J. J. Schneider, R. C. Hoffmann, J. Engstler, O. Soffke, W. Jaegermann, A. Issanin and A. Klyszcz, *Adv. Mater.*, 2008, **20**, 3383-3387.

67. E. Carlos, J. Leppäniemi, A. Sneck, A. Alastalo, J. Deuermeier, R. Branquinho, R. Martins and E. Fortunato, *Adv. Electron. Mater.*, 2020, **6**, 1901071.

68. W. J. Scheideler, M. W. McPhail, R. Kumar, J. Smith and V. Subramanian, *ACS Appl. Mater. Interfaces*, 2018, **10**, 37277-37286.

69. S. Choi, K.-T. Kim, S. K. Park and Y.-H. Kim, *Materials*, 2019, **12**, 852.

70. S. Bolat, P. Fuchs, S. Knobelspies, O. Temel, G. T. Sevilla, E. Gilshtein, C. Andres, I. Shorubalko, Y. Liu and G. Tröster, *Adv. Electron. Mater.*, 2019, **5**, 1800843.

71. C. Avis, H. R. Hwang and J. Jang, *ACS Appl. Mater. Interfaces*, 2014, **6**, 10941-10945.

72. A. H. Adl, P. Kar, S. Farsinezhad, H. Sharma and K. Shankar, *RSC Adv.*, 2015, **5**, 87007-87018.

73. J. Schneider, R. Hoffmann, J. Engstler and O. So, *Adv. Mater*, 2008, **20**, 3383-3387.

74. E. Fortunato, P. Barquinha and R. Martins, *Adv. Mater.*, 2012, **24**, 2945-2986.

75. X. Yu, T. J. Marks and A. Facchetti, *Nat. Mater.*, 2016, **15**, 383-396.

76. J. J. Schneider, R. C. Hoffmann, J. Engstler, S. Dilfer, A. Klyszcz, E. Erdem, P. Jakes and R. A. Eichel, *J. Mater. Chem.*, 2009, **19**, 1449-1457.

77. S. Sanctis, R. C. Hoffmann and J. J. Schneider, *RSC Adv.*, 2013, **3**, 20071-20076.

78. J. J. Schneider, R. C. Hoffmann, A. Issanin and S. Dilfer, *Mater. Sci. Eng. B*, 2011, **176**, 965-971.

79. R. C. Hoffmann, M. Kaloumenos, E. Erdem, S. Weber and J. J. Schneider, *Eur. J. Inorg. Chem.*, 2014, **2014**, 5554-5560.

80. R. C. Hoffmann, M. Kaloumenos, D. Spiehl, E. Erdem, S. Repp, S. Weber and J. J. Schneider, *Phys. Chem. Chem. Phys.*, 2015, **17**, 31801-31809.

81. J. F. Wager, *Science*, 2003, **300**, 1245-1246.

82. G. Thomas, *Nature*, 1997, **389**, 907-908.

83. K. Nomura, H. Ohta, K. Ueda, T. Kamiya, M. Hirano and H. Hosono, *Science*, 2003, **300**, 1269-1272.

84. S. Masuda, K. Kitamura, Y. Okumura, S. Miyatake, H. Tabata and T. Kawai, *J. Appl. Phys.*, 2003, **93**, 1624-1630.

85. Y. Ohya, T. Niwa, T. Ban and Y. Takahashi, *Jpn. J. Appl. Phys.*, 2001, **40**, 297.

86. J. F. Wager, B. Yeh, R. L. Hoffman and D. A. Keszler, *Curr. Opin. Solid State Mater. Sci.*, 2014, **18**, 53-61.

87. V. K. Gueorguiev, L. I. Popova, G. D. Beshkov and N. A. Tomajova, *Sens. Actuators A*, 1990, **24**, 61-63.

88. T. L. Breen, P. M. Fryer, R. W. Nunes and M. E. Rothwell, *Langmuir*, 2002, **18**, 194-197.

89. Y. Shigesato and D. C. Paine, *Appl. Phys. Lett.*, 1993, **62**, 1268-1270.

90. P. Carcia, R. McLean, M. Reilly and G. Nunes Jr, *Appl. Phys. Lett.*, 2003, **82**, 1117-1119.

91. E. Fortunato, A. Pimentel, L. Pereira, A. Gonçalves, G. Lavareda, H. Águas, I. Ferreira, C. N. Carvalho and R. Martins, *J. Non-Cryst. Solids*, 2004, **338-340**, 806-809.

92. T. Kamiya and H. Hosono, *Handbook Vis. Disp. Technol.*, 2014, 1-28.

93. S. J. Kim, S. Yoon and H. J. Kim, *Jpn. J. Appl. Phys.*, 2014, **53**, 02BA02.

94. T. Kawamura, H. Uchiyama, S. Saito, H. Wakana, T. Mine, M. Hatano, K. Torii and T. Onai, *IEEE (IEDM)*, 2008, pp. 1-4.

95. T. Kamiya, K. Nomura and H. Hosono, *J. Disp. Technol.*, 2009, **5**, 468-483.

96. B.-Y. Oh, M.-C. Jeong, M.-H. Ham and J.-M. Myoung, *Semicond. Sci. Technol.*, 2007, **22**, 608.

97. R. Hayashi, M. Ofuji, N. Kaji, K. Takahashi, K. Abe, H. Yabuta, M. Sano, H. Kumomi, K. Nomura and T. Kamiya, *J. Soc. Inf. Displ.*, 2007, **15**, 915-921.

98. R. Martins, A. Nathan, R. Barros, L. Pereira, P. Barquinha, N. Correia, R. Costa, A. Ahnood, I. Ferreira and E. Fortunato, *Adv. Mater.*, 2011, **23**, 4491-4496.

99. X. Yu, J. Smith, N. Zhou, L. Zeng, P. Guo, Y. Xia, A. Alvarez, S. Aghion, H. Lin, J. Yu, R. P. H. Chang, M. J. Bedzyk, R. Ferragut, T. J. Marks and A. Facchetti, *Proc. Natl. Acad. Sci.*, 2015, **112**, 3217-3222.

100. W. F. Smith, J. Hashemi and F. Presuel-Moreno, *Foundations of materials science and engineering*, Mcgraw-Hill Publishing, 2006.

101. M. Ieda, G. Sawa and S. Kato, *J. Appl. Phys.*, 1971, **42**, 3737-3740.

102. S. Chakraborty, M. Bera, S. Bhattacharya and C. Maiti, *Microelectron. Eng.*, 2005, **81**, 188-193.

103. H. Wang, W. Xu, S. Zhou, F. Xie, Y. Xiao, L. Ye, J. Chen and J. Xu, *J. Appl. Phys.*, 2015, **117**, 035703.

104. H. Spahr, S. Montzka, J. Reinker, F. Hirschberg, W. Kowalsky and H.-H. Johannes, *J. Appl. Phys.*, 2013, **114**, 183714.

105. J. R. Macdonald and E. Barsoukov, *History*, 2005, **1**, 1-13.

106. E. Fortunato, P. Barquinha and R. Martins, *Adv. Mater.*, 2012, **24**, 2945-2986.

107. D. K. Schroder, *Semiconductor material and device characterization*, John Wiley & Sons, 2015.

108. S. M. Sze and K. K. Ng, *Physics of semiconductor devices*, John wiley & sons, 2006.

109. R. F. Pierret, *Semiconductor device fundamentals*, Pearson Education India, 1996.

110. G. Horowitz, *Adv. Funct. Mater.*, 2003, **13**, 53-60.

111. W. Boukhili, M. Mahdouani, M. Erouel, J. Puigdollers and R. Bourguiga, *Synth. Met.*, 2015, **199**, 303-309.

112. W. Boukhili, M. Mahdouani, R. Bourguiga and J. Puigdollers, *Superlattices Microstruct.*, 2015, **83**, 224-236.

113. R. P. Ortiz, A. Facchetti and T. J. Marks, *Chem. Rev.*, 2010, **110**, 205-239.

114. A. Liu, H. Zhu, H. Sun, Y. Xu and Y. Y. Noh, *Adv. Mater.*, 2018, **30**, 1706364.

115. D. G. Schlom and J. H. Haeni, *MRS Bull.*, 2002, **27**, 198-204.

116. A. Fissel, H. Osten and E. Bugiel, *J. Vac. Sci. Technol. B*, 2003, **21**, 1765-1772.

117. J. Kwo, M. Hong, A. Kortan, K. Queeney, Y. Chabal, J. Mannaerts, T. Boone, J. Krajewski, A. Sergent and J. Rosamilia, *Appl. Phys. Lett.*, 2000, **77**, 130-132.

118. S. Ohmi, M. Takeda, H. Ishiwara and H. Iwai, *J. Electrochem. Soc.*, 2004, **151**, G279.

119. J. F. Wager, D. A. Keszler and R. E. Presley, *Transparent Electronics*, Springer, New York, 2008.

120. J. Robertson, *J. Vac. Sci. Technol. B*, 2000, **18**, 1785-1791.

121. P. Barquinha, L. Pereira, G. Gonçalves, R. Martins, E. Fortunato, D. Kuscer, M. Kosec, A. Vilà, A. Olziersky and J. R. Morante, *J. Soc. Inf. Disp.*, 2010, **18**, 762-772.

122. J. H. Park, K. Kim, Y. B. Yoo, S. Y. Park, K.-H. Lim, K. H. Lee, H. K. Baik and Y. S. Kim, *J. Mater. Chem. C*, 2013, **1**, 7166-7174.

123. J. Robertson and B. Falabretti, *J. Appl. Phys.*, 2006, **100**, 014111.

124.   J. Robertson and K. Xiong, *Rare Earth Oxide Thin Films*, Springer, 2007, pp. 313-329.

125.   C. Avis, Y. G. Kim and J. Jang, *J. Mater. Chem.*, 2012, **22**, 17415-17420.

126.   F. Zhang, G. Liu, A. Liu, B. Shin and F. Shan, *Ceram. Int.*, 2015, **41**, 13218-13223.

127.   M. Esro, G. Vourlias, C. Somerton, W. I. Milne and G. Adamopoulos, *Adv. Funct. Mater.*, 2015, **25**, 134-141.

128.   L. Xifeng, X. Enlong and Z. Jianhua, *IEEE Trans. Electron. Devices*, 2013, **60**, 3413-3416.

129.   G. Adamopoulos, S. Thomas, D. D. Bradley, M. A. McLachlan and T. D. Anthopoulos, *Appl. Phys. Lett.*, 2011, **98**, 123503.

130.   Y. Xu, X. Li, L. Zhu and J. Zhang, *Mater. Sci. Semicond. Process.*, 2016, **46**, 23-28.

131.   H. Wang, T. Sun, W. Xu, F. Xie, L. Ye, Y. Xiao, Y. Wang, J. Chen and J. Xu, *RSC Adv.*, 2014, **4**, 54729-54739.

132.   H. J. Ha, S. W. Jeong, T.-Y. Oh, M. Kim, K. Choi, J. H. Park and B.-K. Ju, *J. Phys. D Appl. Phys.*, 2013, **46**, 235102.

133.   E. J. Bae, Y. H. Kang, M. Han, C. Lee and S. Y. Cho, *J. Mater. Chem. C*, 2014, **2**, 5695-5703.

134.   W. Xu, M. Long, T. Zhang, L. Liang, H. Cao, D. Zhu and J.-B. Xu, *Ceram. Int.*, 2017, **43**, 6130-6137.

135.   H. Tan, G. Liu, A. Liu, B. Shin and F. Shan, *Ceram. Int.*, 2015, **41**, S349-S355.

136.   H. Park, Y. Nam, J. Jin and B.-S. Bae, *RSC Adv.*, 2015, **5**, 102362-102366.

137.   Y. S. Rim, H. Chen, T.-B. Song, S.-H. Bae and Y. Yang, *Chem. Mater.*, 2015, **27**, 5808-5812.

138.   Y. Zhang, G. Huang, L. Duan, G. Dong, D. Zhang and Y. Qiu, *Sci. China Technol. Sci.*, 2016, **59**, 1407-1412.

139.   Y. Gao, Y. Xu, J. Lu, J. Zhang and X. Li, *J. Mater. Chem. C*, 2015, **3**, 11497-11504.

140.   X. Li, L. Zhu, Y. Gao and J. Zhang, *IEEE Trans. Electron. Devices*, 2015, **62**, 875-881.

141.   I. Krylov, D. Ritter and M. Eizenberg, *J. Appl. Phys.*, 2017, **122**, 034505.

142.   G. He, W. Li, Z. Sun, M. Zhang and X. Chen, *RSC Adv.*, 2018, **8**, 36584-36595.

143.   L. Zhu, Y. Gao, X. Li, X. Sun and J. Zhang, *J. Mater. Res.*, 2014, **29**, 1620-1625.

144.   M. Bizarro, J. Alonso and A. Ortiz, *Mater. Sci. Semicond. Process.*, 2006, **9**, 1090-1096.

145.   W. Yang, K. Song, Y. Jung, S. Jeong and J. Moon, *J. Mater. Chem. C*, 2013, **1**, 4275-4282.

146.   K. Woods, E. Waddington, C. Crump, E. Bryan, T. Gleckler, M. Nellist, B. Duell, D. Nguyen, S. Boettcher and C. Page, *RSC Adv.*, 2017, **7**, 39147-39152.

147.   L. Zhu, G. He, C. Zhang, B. Yang, Y. Xia, F. Alam and Y. Zhang, *IEEE Trans. Electron. Devices*, 2019, **66**, 4198-4204.

148.   P. N. Plassmeyer, K. Archila, J. F. Wager and C. J. Page, *ACS Appl. Mater. Interfaces*, 2015, **7**, 1678-1684.

149.   J. Lee, H. Seul and J. K. Jeong, *J. Alloys Compd.*, 2018, **741**.

150. S. Bolat, P. Fuchs, S. Knobelspies, O. Temel, G. T. Sevilla, E. Gilshtein, C. Andres, I. Shorubalko, Y. Liu and G. Tröster, *Adv. Electron. Mater.*, 2019, 1800843.

151. A. Mancinelli, S. Bolat, J. Kim, Y. E. Romanyuk and D. Briand, *ACS Appl. Electron. Mater.*, 2020.

152. G. Teowee, K. McCarthy, F. McCarthy, T. Bukowski, D. Davis and D. R. Uhlmann, *J. Solgel Sci. Technol.*, 1998, **13**, 895-898.

153. H. Zhang, L. Liang, A. Chen, Z. Liu, Z. Yu, H. Cao and Q. Wan, *Appl. Phys. Lett.*, 2010, **97**, 122108.

154. X. Huang and P. Lai, in *2010 IEEE (EDSSC)*, 2010, pp. 1-4.

155. P. Katiyar, C. Jin and R. Narayan, *Acta Mater.*, 2005, **53**, 2617-2622.

156. S. Yaginuma, J. Yamaguchi, K. Itaka and H. Koinuma, *Thin Solid Films*, 2005, **486**, 218-221.

157. M. Ion, C. Berbecaru, S. Iftimie, M. Filipescu, M. Dinescu and S. Antohe, *Dig. J. Nanomater. Bios.*, 2012, **7**, 1609-1614.

158. J. Zhu and Z. Liu, *Microelectron. Eng.*, 2003, **66**, 849-854.

159. A. Suresh, P. Gollakota, P. Wellenius, A. Dhawan and J. F. Muth, *Thin Solid Films*, 2008, **516**, 1326-1329.

160. R. M. Pasquarelli, D. S. Ginley and R. O'Hayre, *Chem. Soc. Rev.*, 2011, **40**, 5406-5441.

161. L. L. Hench and J. K. West, *Chem. Rev.*, 1990, **90**, 33-72.

162. B. Wang, X. Yu, P. Guo, W. Huang, L. Zeng, N. Zhou, L. Chi, M. J. Bedzyk, R. P. Chang and T. J. Marks, *Adv. Electron. Mater.*, 2016, **2**, 1500427.

163. B. Sun and H. Sirringhaus, *Nano letters*, 2005, **5**, 2408-2413.

164. S. Lee, Y. Jeong, S. Jeong, J. Lee, M. Jeon and J. Moon, *Superlattices Microstruct.*, 2008, **44**, 761-769.

165. C. S. Li, Y. N. Li, Y. L. Wu, B. S. Ong and R. O. Loutfy, *J. Phys. D Appl. Phys.*, 2008, **41**, 125102.

166. V. Subramanian, T. Bakhishev, D. Redinger and S. K. Volkman, *J. Disp. Technol.*, 2009, **5**, 525-530.

167. R. Theissmann, S. Bubel, M. Sanlialp, C. Busch, G. Schierning and R. Schmechel, *Thin Solid Films*, 2011, **519**, 5623-5628.

168. K. Choi, M. Kim, S. Chang, T.-Y. Oh, S. W. Jeong, H. J. Ha and B.-K. Ju, *Jpn. J. Appl. Phys.*, 2013, **52**, 060204.

169. H. S. Kim, P. D. Byrne, A. Facchetti and T. J. Marks, *Journal of the American Chemical Society*, 2008, **130**, 12580-12581.

170. S.-Y. Han, G. S. Herman and C.-h. Chang, *J. Am. Chem. Soc.*, 2011, **133**, 5166-5169.

171. Y. Meng, G. Liu, A. Liu, H. Song, Y. Hou, B. Shin and F. Shan, *RSC Adv.*, 2015, **5**, 37807-37813.

172. G. Huang, L. Duan, G. Dong, D. Zhang and Y. Qiu, *ACS Appl. Mater. Interfaces*, 2014, **6**, 20786-20794.

173. H. S. Kim, M.-G. Kim, Y.-G. Ha, M. G. Kanatzidis, T. J. Marks and A. Facchetti, *J. Am. Chem. Soc.*, 2009, **131**, 10826-10827.

174. C. G. Choi, S.-J. Seo and B.-S. Bae, *ECS Solid State Lett.*, 2007, **11**, H7.

175. C. Y. Koo, K. Song, T. Jun, D. Kim, Y. Jeong, S.-H. Kim, J. Ha and J. Moon, *J. Electrochem. Soc.*, 2010, **157**, J111.

176. S. Jeong, Y. Jeong and J. Moon, *J. Phys. Chem. C*, 2008, **112**, 11082-11085.

177. S.-J. Seo, C. G. Choi, Y. H. Hwang and B.-S. Bae, *J. Phys. D Appl. Phys.*, 2008, **42**, 035106.

178. C.-G. Lee and A. Dodabalapur, *Appl. Phys. Lett.*, 2010, **96**, 243501.

179. M.-G. Kim, H. S. Kim, Y.-G. Ha, J. He, M. G. Kanatzidis, A. Facchetti and T. J. Marks, *J. Am. Chem. Soc.*, 2010, **132**, 10352-10364.

180. G. H. Kim, H. S. Shin, B. Du Ahn, K. H. Kim, W. J. Park and H. J. Kim, *J. Electrochem. Soc.*, 2008, **156**, H7.

181. G. H. Kim, B. Du Ahn, H. S. Shin, W. H. Jeong, H. J. Kim and H. J. Kim, *Appl. Phys. Lett.*, 2009, **94**, 233501.

182. S. Jeong, Y. G. Ha, J. Moon, A. Facchetti and T. J. Marks, *Adv. Mater.*, 2010, **22**, 1346-1350.

183. S. Hwang, J. H. Lee, C. H. Woo, J. Y. Lee and H. K. Cho, *Thin Solid Films*, 2011, **519**, 5146-5149.

184. W. Xu, L. Hu, C. Zhao, L. Zhang, D. Zhu, P. Cao, W. Liu, S. Han, X. Liu and F. Jia, *Appl. Surf. Sci.*, 2018, **455**, 554-560.

185. F. Babonneau, S. Doeuff, A. Leaustic, C. Sanchez, C. Cartier and M. Verdaguer, *Inorg. Chem.*, 1988, **27**, 3166-3172.

186. U. Schubert, *J. Mater. Chem.*, 2005, **15**, 3701-3715.

187. M. Yu, M. Liang, J. Liu, S. Li, B. Xue and H. Zhao, *Appl. Surf. Sci.*, 2016, **363**, 229-239.

188. Q.-J. Sun, J. Peng, W.-H. Chen, X.-J. She, J. Liu, X. Gao, W.-L. Ma and S.-D. Wang, *Org. Electron.*, 2016, **34**, 118-123.

189. D. E. Walker, M. Major, M. Baghaie Yazdi, A. Klyszcz, M. Haeming, K. Bonrad, C. Melzer, W. Donner and H. von Seggern, *ACS Appl. Mater. Interfaces*, 2012, **4**, 6835-6841.

190. S.-Y. Han, D.-H. Lee, G. S. Herman and C.-H. Chang, *J. Disp. Technol.*, 2009, **5**, 520-524.

191. Y. Zhao, G. Dong, L. Duan, J. Qiao, D. Zhang, L. Wang and Y. Qiu, *RSC Adv.*, 2012, **2**, 5307-5313.

192. B. Sykora, D. Wang and H. von Seggern, *Appl. Phys. Lett.*, 2016, **109**, 033501.

193. D. H. Lee, S. M. Park, J. I. Yang, D. K. Cho, S. H. Woo, Y. S. Lim, D. K. Kim and M. Yi, *Jpn. J. Appl. Phys.*, 2013, **52**, 10MA02.

194. S. Jeong and J. Moon, *J. Mater. Chem.*, 2012, **22**, 1243-1250.

195. S. W. Smith, W. Wang, D. A. Keszler and J. F. Conley Jr, *J. Vac. Sci. Technol. A*, 2014, **32**, 041501.

196. W. H. Jeong, J. H. Bae and H. J. Kim, *IEEE Electron Device Lett.*, 2011, **33**, 68-70.

197. J. r. J. Schneider, R. C. Hoffmann, J. r. Engstler, A. Klyszcz, E. Erdem, P. Jakes, R. d.-A. Eichel, L. Pitta-Bauermann and J. Bill, *Chem. Mater.*, 2010, **22**, 2203-2212.

198. R. C. Hoffmann, S. Dilfer, A. Issanin and J. J. Schneider, *Phys. Status Solidi A*, 2010, **207**, 1590-1595.

199. R. C. Hoffmann, M. Kaloumenos, S. Heinschke, E. Erdem, P. Jakes, R.-A. Eichel and J. J. Schneider, *J. Mater. Chem. C*, 2013, **1**, 2577-2584.

200. S. Sanctis, R. C. Hoffmann, S. Eiben and J. J. Schneider, *Beilstein J. Nanotechnol.*, 2015, **6**, 785-791.

201. Y.-J. Chang, D.-H. Lee, G. Herman and C.-H. Chang, *ECS Solid State Lett.*, 2007, **10**, H135.

202. H. Chiang, J. Wager, R. Hoffman, J. Jeong and D. A. Keszler, *Appl. Phys. Lett.*, 2005, **86**, 013503.

203. L.-C. Liu, J.-S. Chen and J.-S. Jeng, *Appl. Phys. Lett.*, 2014, **105**, 023509.

204. S.-P. Tsai, C.-H. Chang, C.-J. Hsu, C.-C. Hu, Y.-T. Tsai, C.-H. Chou, H.-H. Lin and C.-C. Wu, *ECS J. Solid State Sci. Technol.*, 2015, **4**, P176.

205. R. D. Chandra, M. Rao, K. Zhang, R. R. Prabhakar, C. Shi, J. Zhang, S. G. Mhaisalkar and N. Mathews, *ACS Appl. Mater. Interfaces*, 2014, **6**, 773-777.

206. S. Sanctis, N. Koslowski, R. Hoffmann, C. Guhl, E. Erdem, S. Weber and J. r. J. Schneider, *ACS Appl. Mater. Interfaces*, 2017, **9**, 21328-21337.

207. S. Sanctis, R. C. Hoffmann, M. Bruns and J. J. Schneider, *Adv. Mater. Interfaces*, 2018, **5**, 1800324.

208. S. Sanctis, R. C. Hoffmann, R. Precht, W. Anwand and J. J. Schneider, *J. Mater. Chem. C*, 2016, **4**, 10935-10944.

209. W. H. Jeong, D. L. Kim and H. J. Kim, *ACS Appl. Mater. Interfaces*, 2013, **5**, 9051-9056.

210. Y. Wang, S. W. Liu, X. W. Sun, J. L. Zhao, G. K. L. Goh, Q. V. Vu and H. Y. Yu, *J. Solgel Sci. Technol.*, 2010, **55**, 322-327.

211. R. C. Hoffmann and J. J. Schneider, *Eur. J. Inorg. Chem.*, 2014, **2014**, 2241-2247.

212. Y. Chen, B. Wang, W. Huang, X. Zhang, G. Wang, M. J. Leonardi, Y. Huang, Z. Lu, T. J. Marks and A. Facchetti, *Chem. Mater.*, 2018, **30**, 3323-3329.

213. N. Koslowski, S. Sanctis, R. C. Hoffmann, M. Bruns and J. J. Schneider, *J. Mater. Chem. C*, 2019, **7**, 1048-1056.

214. N. Koslowski, R. C. Hoffmann, V. Trouillet, M. Bruns, S. Foro and J. J. Schneider, *RSC Adv.*, 2019, **9**, 31386-31397.

215. N. Koslowski, V. Trouillet and J. J. Schneider, *J. Mater. Chem. C*, 2020.

216. S. Aruna and A. Mukasyan, 2008.

217. A. Varma, A. S. Mukasyan, A. S. Rogachev and K. V. Manukyan, *Chem. Rev.*, 2016, **116**, 14493-14586.

218. M. Lupin and G. Peters, *Thermochim. Acta*, 1984, **73**, 79-87.

219. S. Désilets, P. Brousseau, D. Chamberland, S. Singh, H. Feng, R. Turcotte, K. Armstrong and J. Anderson, *Thermochim. Acta*, 2011, **521**, 59-65.

220. M.-G. Kim, M. G. Kanatzidis, A. Facchetti and T. J. Marks, *Nat. Mater.*, 2011, **10**, 382-388.

221. K. Banger, Y. Yamashita, K. Mori, R. Peterson, T. Leedham, J. Rickard and H. Sirringhaus, *Nat. Mater.*, 2011, **10**, 45-50.

222. J. W. Hennek, M.-G. Kim, M. G. Kanatzidis, A. Facchetti and T. J. Marks, *J. Am. Chem. Soc.*, 2012, **134**, 9593-9596.

223. J. W. Hennek, J. Smith, A. Yan, M.-G. Kim, W. Zhao, V. P. Dravid, A. Facchetti and T. J. Marks, *J. Am. Chem. Soc.*, 2013, **135**, 10729-10741.

224. X. Yu, N. Zhou, J. Smith, H. Lin, K. Stallings, J. Yu, T. J. Marks and A. Facchetti, *ACS Appl. Mater. Interfaces*, 2013, **5**, 7983-7988.

225. G. Liu, A. Liu, F. Shan, Y. Meng, B. Shin, E. Fortunato and R. Martins, *Appl. Phys. Lett.*, 2014, **105**, 113509.

226. H. Aghabozorg, G. J. Palenik, R. C. Stoufer and J. Summers, *Inorg. Chem.*, 1982, **21**, 3903-3907.

227. B. Figgis, E. Kucharski, J. Patrick and A. White, *Aust. J. Chem.*, 1984, **37**, 265-271.

228. D. Moon, S. Tanaka, T. Akitsu and J.-H. Choi, *Acta Cryst. E*, 2015, **71**, 1336-1339.

229. Y. Qiu and L. Gao, *J. Am. Ceram. Soc.*, 2004, **87**, 352-357.

230. T. J. Prior and R. L. Kift, *J. Chem. Crystallogr. - Cryst. Mater.*, 2009, **39**, 558-563.

231. A. Krawczuk and K. Stadnicka, *Acta Cryst. C*, 2007, **63**, m448-m450.

232. M. Koman, E. Jona and D. Nagy, *Z. Kristallogr.*, 1995, **210**, 873-874.

233. S. Smeets and M. Lutz, *Acta Cryst. C*, 2011, **67**, m50-m55.

234. S. Sanctis, R. C. Hoffmann, N. Koslowski, S. Foro, M. Bruns and J. J. Schneider, *Chem. Asian J.*, 2018, **13**, 3912-3919.

235. W. S. Wong and A. Salleo, *Flexible electronics: materials and applications*, Springer Science & Business Media, 2009.

236. Y. M. Park, J. Daniel, M. Heeney and A. Salleo, *Adv. Mater.*, 2011, **23**, 971-974.

237. Y.-H. Kim, J.-S. Heo, T.-H. Kim, S. Park, M.-H. Yoon, J. Kim, M. S. Oh, G.-R. Yi, Y.-Y. Noh and S. K. Park, *Nature*, 2012, **489**, 128-132.

238. Y.-H. Lin, H. Faber, K. Zhao, Q. Wang, A. Amassian, M. McLachlan and T. D. Anthopoulos, *Adv. Mater.*, 2013, **25**, 4340-4346.

239. J. S. Heo, J.-W. Jo, J. Kang, C.-Y. Jeong, H. Y. Jeong, S. K. Kim, K. Kim, H.-I. Kwon, J. Kim and Y.-H. Kim, *ACS Appl. Mater. Interfaces*, 2016, **8**, 10403-10412.

240. S. Park, K. H. Kim, J. W. Jo, S. Sung, K. T. Kim, W. J. Lee, J. Kim, H. J. Kim, G. R. Yi and Y. H. Kim, *Adv. Funct. Mater.*, 2015, **25**, 2807-2815.

241. I. Bretos, R. Jiménez, J. Ricote and M. L. Calzada, *Chem. Eur. J.*, 2020.

242. E. Carlos, R. Branquinho, A. Kiazadeh, P. Barquinha, R. Martins and E. Fortunato, *ACS Appl. Mater. Interfaces*, 2016, **8**, 31100-31108.

243. R. A. John, N. A. Chien, S. Shukla, N. Tiwari, C. Shi, N. G. Ing and N. Mathews, *Chem. Mater.*, 2016, **28**, 8305-8313.

244. G. X. Liu, A. Liu, F. K. Shan, Y. Meng, B. C. Shin, E. Fortunato and R. Martins, *Appl. Phys. Lett.*, 2014, **105**, 113509.

245. A. Albini, *Photochem. Photobiol. Sci.*, 2016, **15**, 319-324.

246. B. R. Strohmeier, *Surf. Sci. Spectra*, 1994, **3**, 135-140.

247. V. Trouillet, H. Troesse, M. Bruns, E. Nold and R. White, *J. Vac. Sci. Technol. A*, 2007, **25**, 927-931.

248. C. Gao, X.-Y. Yu, R.-X. Xu, J.-H. Liu and X.-J. Huang, *ACS Appl. Mater. Interfaces*, 2012, **4**, 4672-4682.

249. H. Oh, J. Krantz, I. Litzov, T. Stubhan, L. Pinna and C. J. Brabec, *Sol. Energy Mater. Sol. Cells*, 2011, **95**, 2194-2199.

250. S. Chen, J. R. Manders, S.-W. Tsang and F. So, *J. Mater. Chem.*, 2012, **22**, 24202-24212.

251. F. Wang, Z. a. Tan and Y. Li, *Energy Environ. Sci.*, 2015, **8**, 1059-1091.

252. L. G. Gerling, S. Mahato, A. Morales-Vilches, G. Masmitja, P. Ortega, C. Voz, R. Alcubilla and J. Puigdollers, *Sol. Energy Mater. Sol. Cells*, 2016, **145**, 109-115.

253. M. A. Haque, A. D. Sheikh, X. Guan and T. Wu, *Adv. Energy Mater.*, 2017, **7**, 1602803.

254. S. Choopun, A. Tubtimtae, T. Santhaveesuk, S. Nilphai, E. Wongrat and N. Hongsith, *Appl. Surf. Sci.*, 2009, **256**, 998-1002.

255. K. Wetchakun, T. Samerjai, N. Tamaekong, C. Liewhiran, C. Siriwong, V. Kruefu, A. Wisitsoraat, A. Tuantranont and S. Phanichphant, *Sens. Actuator A Phys.*, 2011, **160**, 580-591.

256. R. Kumar, X. Liu, J. Zhang and M. Kumar, *Nano-Micro Lett.*, 2020, **12**, 1-37.